NMTA 25 Middle Lev
Science

Teacher Certification Exam

By: Sharon Wynne, M.S
Southern Connecticut State University

XAMonline, INC.
Boston

Copyright © 2007 XAMonline, Inc.
All rights reserved. No part of the material protected by this copyright notice may be reproduced or utilized in any form or by any means, electronic or mechanical, including photocopying, recording or by any information storage and retrievable system, without written permission from the copyright holder.

To obtain permission(s) to use the material from this work for any purpose including workshops or seminars, please submit a written request to:

XAMonline, Inc.
21 Orient Ave.
Melrose, MA 02176
Toll Free 1-800-509-4128
Email: info@xamonline.com
Web www.xamonline.com
Fax: 1-781-662-9268

Library of Congress Cataloging-in-Publication Data

Wynne, Sharon A.
Middle Level Science 25: Teacher Certification / Sharon A. Wynne. -2nd ed.
ISBN 978-1-58197-758-5
1. Middle Level Science 25. 2. Study Guides. 3. NMTA
4. Teachers' Certification & Licensure. 5. Careers

Disclaimer:
The opinions expressed in this publication are the sole works of XAMonline and were created independently from the National Education Association, Educational Testing Service, or any State Department of Education, National Evaluation Systems or other testing affiliates.

Between the time of publication and printing, state specific standards as well as testing formats and website information may change that is not included in part or in whole within this product. Sample test questions are developed by XAMonline and reflect similar content as on real tests; however, they are not former tests. XAMonline assembles content that aligns with state standards but makes no claims nor guarantees teacher candidates a passing score. Numerical scores are determined by testing companies such as NES or ETS and then are compared with individual state standards. A passing score varies from state to state.

Printed in the United States of America **œ-1**

NMTA: Middle Level Science 25
ISBN: 978-1-58197-758-5

Table of Contents

SUBAREA II. LIFE SCIENCE

SUBAREA III. PHYSICAL SCIENCE

SUBAREA IV. EARTH AND SPACE SCIENCE

Great Study and Testing Tips!

What to study in order to prepare for the subject assessments is the focus of this study guide but equally important is how you study.

You can increase your chances of truly mastering the information by taking some simple, but effective steps.

Study Tips:

1. **Some foods aid the learning process.** Foods such as milk, nuts, seeds, rice, and oats help your study efforts by releasing natural memory enhancers called CCKs (cholecystokinin) composed of tryptophan, choline, and phenylalanine. All of these chemicals enhance the neurotransmitters associated with memory. Before studying, try a light, protein-rich meal of eggs, turkey, and fish. All of these foods release the memory enhancing chemicals. The better the connections, the more you comprehend.

Likewise, before you take a test, stick to a light snack of energy boosting and relaxing foods. A glass of milk, a piece of fruit, or some peanuts all release various memory-boosting chemicals and help you to relax and focus on the subject at hand.

2. **Learn to take great notes.** A by-product of our modern culture is that we have grown accustomed to getting our information in short doses (i.e. TV news sound bites or USA Today style newspaper articles.)

Consequently, we've subconsciously trained ourselves to assimilate information better in neat little packages. If your notes are scrawled all over the paper, it fragments the flow of the information. Strive for clarity. Newspapers use a standard format to achieve clarity. Your notes can be much clearer through use of proper formatting. A very effective format is called the *"Cornell Method."*

> Take a sheet of loose-leaf lined notebook paper and draw a line all the way down the paper about 1-2" from the left-hand edge.
>
> Draw another line across the width of the paper about 1-2" up from the bottom. Repeat this process on the reverse side of the page.

Look at the highly effective result. You have ample room for notes, a left hand margin for special emphasis items or inserting supplementary data from the textbook, a large area at the bottom for a brief summary, and a little rectangular space for just about anything you want.

3. **Get the concept then the details.** Too often we focus on the details and don't gather an understanding of the concept. However, if you simply memorize only dates, places, or names, you may well miss the whole point of the subject.

A key way to understand things is to put them in your own words. If you are working from a textbook, automatically summarize each paragraph in your mind. If you are outlining text, don't simply copy the author's words.

Rephrase them in your own words. You remember your own thoughts and words much better than someone else's, and subconsciously tend to associate the important details to the core concepts.

4. **Ask Why?** Pull apart written material paragraph by paragraph and don't forget the captions under the illustrations.

Example: If the heading is "Stream Erosion", flip it around to read "Why do streams erode?" Then answer the questions.

If you train your mind to think in a series of questions and answers, not only will you learn more, but it also helps to lessen the test anxiety because you are used to answering questions.

5. **Read for reinforcement and future needs.** Even if you only have 10 minutes, put your notes or a book in your hand. Your mind is similar to a computer; you have to input data in order to have it processed. By reading, you are creating the neural connections for future retrieval. The more times you read something, the more you reinforce the learning of ideas.

Even if you don't fully understand something on the first pass, your mind stores much of the material for later recall.

6. **Relax to learn so go into exile.** Our bodies respond to an inner clock called biorhythms. Burning the midnight oil works well for some people, but not everyone.

If possible, set aside a particular place to study that is free of distractions. Shut off the television, cell phone, pager and exile your friends and family during your study period.

If you really are bothered by silence, try background music. Light classical music at a low volume has been shown to aid in concentration over other types. Music that evokes pleasant emotions without lyrics are highly suggested. Try just about anything by Mozart. It relaxes you.

7. Use arrows not highlighters. At best, it's difficult to read a page full of yellow, pink, blue, and green streaks. Try staring at a neon sign for a while and you'll soon see that the horde of colors obscure the message.

A quick note, a brief dash of color, an underline, and an arrow pointing to a particular passage is much clearer than a horde of highlighted words.

8. Budget your study time. Although you shouldn't ignore any of the material, **ALLOCATE YOUR AVAILABLE STUDY TIME IN THE SAME RATIO THAT TOPICS MAY APPEAR ON THE TEST.**

Testing Tips:

1. **Get smart, play dumb. Don't read anything into the question.** Don't make an assumption that the test writer is looking for something else than what is asked. Stick to the question as written and don't read extra things into it.

2. **Read the question and all the choices *twice* before answering the question.** You may miss something by not carefully reading, and then re-reading both the question and the answers.

If you really don't have a clue as to the right answer, leave it blank on the first time through. Go on to the other questions, as they may provide a clue as to how to answer the skipped questions.

If later on, you still can't answer the skipped ones . . . **GUESS.** The only penalty for guessing is that you might get it wrong. Only one thing is certain; if you don't put anything down, you will get it wrong!

3. **Turn the question into a statement.** Look at the way the questions are worded. The syntax of the question usually provides a clue. Does it seem more familiar as a statement rather than as a question? Does it sound strange?

By turning a question into a statement, you may be able to spot if an answer sounds right, and it may also trigger memories of material you have read.

4. **Look for hidden clues.** It's actually very difficult to compose multiple-foil (choice) questions without giving away part of the answer in the options presented.

In most multiple-choice questions you can often readily eliminate one or two of the potential answers. This leaves you with only two real possibilities and automatically your odds go to Fifty-Fifty for very little work.

5. **Trust your instincts.** For every fact that you have read, you subconsciously retain something of that knowledge. On questions that you aren't really certain about, go with your basic instincts. **Your first impression on how to answer a question is usually correct.**

6. **Mark your answers directly on the test booklet.** Don't bother trying to fill in the optical scan sheet on the first pass through the test.

Just be very careful not to miss-mark your answers when you eventually transcribe them to the scan sheet.

7. **Watch the clock!** You have a set amount of time to answer the questions. Don't get bogged down trying to answer a single question at the expense of 10 questions you can more readily answer.

SUBAREA I. SCIENTIFIC INQUIRY AND THE HISTORY AND NATURE OF SCIENCE

COMPETENCY 1.0 UNDERSTAND THE NATURE OF SCIENCE AND THE PROCESSES AND PRINCIPLES OF SCIENTIFIC INQUIRY

Skill 1.1 Demonstrate an understanding of the nature, purpose, and distinguishing characteristics of science (e.g., the use of empirical standards, logical argument, skepticism) and limitations to the scope of science

The combination of science, mathematics and technology forms the scientific endeavor and makes science a success. It is impossible to study science on its own without the support of other disciplines like mathematics, technology, geology, physics, and other disciplines as well.

Science is tentative. By definition it is searching for information by making educated guesses. It must be replicable. Another scientist must be able to achieve the same results under the same conditions at a later time. The term empirical means it must be assessed through tests and observations. Science changes over time. Science is limited by the available technology. An example of this would be the relationship of the discovery of the cell and the invention of the microscope. As our technology improves, more hypotheses will become theories and possibly laws. Science is also limited by the data that is able to be collected. Data may be interpreted differently on different occasions. Science limitations cause explanations to be changeable as new technologies emerge. New technologies gather previously unavailable data and enable us to build upon current theories with new information.

Ancient history followed the geocentric theory, which was displaced by the heliocentric theory developed by Copernicus, Ptolemy and Kepler. Newton's laws of motion by Sir Isaac Newton were based on mass, force and acceleration, and state that the force of gravity between any two objects in the universe depends upon their mass and distance. These laws are still widely used today. In the 20th century, Albert Einstein was the most outstanding scientist for his work on relativity, which led to his theory that E=mc2. Early in the 20th century, Alfred Wegener proposed his theory of continental drift, stating that continents moved away from the super continent, Pangaea. This theory was accepted in 1960s when more evidence was collected on this. John Dalton and Lavosier made significant contributions in the field of the atom and matter. The Curies and Ernest Rutherford contributed greatly to radioactivity and the splitting of the atom, which have a lot of practical applications. Charles Darwin proposed his theory of evolution and Gregor Mendel's experiments on peas helped us to understand heredity. The most significant improvement was the Industrial Revolution in Britain, in which science was applied practically to increase the productivity and also introduced a number of social problems like child labor.

The nature of science mainly consists of three important things:

1. The scientific world view

This includes some very important issues like – it is possible to understand this highly organized world and its complexities with the help of latest the technology. Scientific ideas are subject to change. After repeated experiments, a theory is established, but this theory could be changed or supported in the future. Only laws that occur naturally do not change. Scientific knowledge may not be discarded but is modified – e.g., Albert Einstein didn't discard the Newtonian principles but modified them in his theory of relativity. Also, science can't answer all of our questions. We can't find answers to questions related to our beliefs, moral values, and our norms.

2. Scientific inquiry

Scientific inquiry starts with a simple question. This simple question leads to information gathering and an educated guess, otherwise known as a hypothesis. To prove the hypothesis, an experiment has to be conducted, which yields data and the conclusion. All experiments must be repeated at least twice to get reliable results. Thus scientific inquiry leads to new knowledge or verifying established theories. Science requires proof or evidence. Science is dependent on accuracy not bias or prejudice. In science, there is no place for preconceived ideas or premeditated results. By using their senses and modern technology, scientists will be able to get reliable information. Science is a combination of logic and imagination. A scientist needs to think and imagine and be able to reason.

Science explains, reasons and predicts. These three are interwoven and are inseparable. While reasoning is absolutely important for science, there should be no bias or prejudice. Science is not authoritarian because it has been shown that scientific authority can be wrong. No one can determine or make decisions for others on any issue.

3. Scientific enterprise

Science is a complex activity involving various people and places. A scientist may work alone or in a laboratory, classroom or for that matter anywhere. Mostly it is a group activity requiring a lot of social skills of cooperation, communication of results or findings, consultations and discussions. Science demands a high degree of communication to the governments, funding authorities and to the public.

Bias

Scientific research can be biased in the choice of what data to consider, in the reporting or recording of the data, and/or in how the data are interpreted. The scientist's emphasis may be influenced by his/her nationality, sex, ethnic origin, age, or political convictions. For example, when studying a group of animals, male scientists may focus on the social behavior of the males and typically male characteristics.

Although bias related to the investigator, the sample, the method, or the instrument may not be completely avoidable in every case, it is important to know the possible sources of bias and how bias could affect the evidence. Moreover, scientists need to be attentive to possible bias in their own work as well as that of other scientists.

Objectivity may not always be attained. However, one precaution that may be taken to guard against undetected bias is to have many different investigators or groups of investigators working on a project. By different, it is meant that the groups are made up of various nationalities, ethnic origins, ages, and political convictions and composed of both males and females. It is also important to note one's aspirations, and to make sure to be truthful to the data, even when grants, promotions, and notoriety are at risk.

Skill 1.2 Recognize the importance of verifiable evidence and peer review in science and that scientific hypotheses are subject to experimental and observational confirmation

Science is a process of checks and balances. It is expected that scientific findings will be challenged, and in many cases retested. Often one experiment will be the beginning point for another. While bias does exist, the use of controlled experiments and an awareness on the part of the scientist, can go far in ensuring a sound experiment. Even if the science is well done, it may still be questioned. It is through this continual search that hypotheses are made into theories, and sometimes become laws. It is also through this search that new information is discovered.

Skill 1.3 Demonstrate knowledge of scientific methods and apply principles and procedures for designing and conducting scientific investigations (e.g., identifying constants, manipulated and responding variables; sampling)

The scientific method is the basic process behind science. It involves several steps beginning with hypothesis formulation and working through to the conclusion.

Posing a question
Although many discoveries happen by chance, the standard thought process of a scientist begins with forming a question to research. The more limited the question, the easier it is to set up an experiment to answer it.

Form a hypothesis
Once the question is formulated take an educated guess about the answer to the problem or question. This 'best guess' is your hypothesis.

Conducting the test
To make a test fair, data from an experiment must have a variable or any condition that can be changed such as temperature or mass. A good test will try to manipulate as few variables as possible so as to see which variable is responsible for the result. This requires a second example of a control. A control is an extra setup in which all the conditions are the same except for the variable being tested.

Observe and record the data
Reporting of the data should state specifics of how the measurements were calculated. A graduated cylinder needs to be read with proper procedures. As beginning students, technique must be part of the instructional process so as to give validity to the data.

Drawing a conclusion
After recording data, you compare your data with that of other groups. A conclusion is the judgment derived from the data results.

Graphing data
Graphing utilizes numbers to demonstrate patterns. The patterns offer a visual representation, making it easier to draw conclusions.

Normally, knowledge is integrated in the form of a lab report. A report has many sections. It should include a specific title and tell exactly what is being studied. The abstract is a summary of the report written at the beginning of the paper. The purpose should always be defined and will state the problem. The purpose should include the hypothesis (educated guess) of what is expected from the outcome of the experiment. The entire experiment should relate to this problem. It is important to describe exactly what was done to prove or disprove a hypothesis. A <u>control</u> is necessary to prove that the results occurred from the changed conditions and would not have happened normally. Only one <u>variable</u> should be manipulated at a time. Observations and results of the experiment should be recorded including data from all results. Drawings, graphs and illustrations should be included to support information. Observations are objective, whereas analysis and interpretation is subjective. A conclusion should explain why the results of the experiment either proved or disproved the hypothesis. A scientific theory is an explanation of a set of related observations based on a proven hypothesis. A scientific law usually lasts longer than a scientific theory and has more experimental data to support it.

Sampling

Sampling is collecting pieces/specimens or making instrument data points/observations at determined intervals or areas for the purpose of research/investigation. Sampling includes animal tracking, capturing, plant and animal tagging, plot sampling, specimen collecting, transect sampling, water sampling etc. The results obtained are used as representative of the whole research area or population.

Skill 1.4 Identify the characteristics and uses of various types of scientific investigations (e.g., controlled experiments, field observations) and evaluate the appropriateness of a given investigative design for testing a particular hypothesis

Scientific investigations come in all sizes and forms. One can conduct a simple survey, over the course of a large population, with the hopes of gaining an understanding of the entire population. This method is often used by medical and pharmaceutical companies and may include a questionnaire that asks about health and lifestyle. Ecologists use field observations. Like the medical questionnaire, they study small sample sizes to gain a better understanding of a larger group. For example, they may track one animal to follow its migratory patterns, or they may place cameras in one area in the hopes of capturing footage of a roaming animal or pack. Ecologists are studying an area and all of the organisms within it, but this is too broad to study, so they often limit sampling size and use a representative of the population. Whenever possible, a scientist would prefer to use controlled experiments. This can happen most readily in a laboratory, and is near impossible to achieve in nature. In a controlled experiment, only one variable is manipulated at once, and a control, or normal variable under normal conditions, is always present. This control group gives the scientist something to compare the variable against. It tells him what would normally have happened under the experimental conditions, had he not altered/introduced the variable.

An experiment is proposed and performed with the sole objective of testing a hypothesis. When evaluating an experiment, it is important to first look at the question it was supposed to answer. How logically did the experiment flow from there? How many variables existed? (it is best to only test one variable at a time) You discover a scientist conducting an experiment with the following characteristics. He has two rows each set up with four stations. The first row has a piece of tile as the base at each station. The second row has a piece of linoleum as the base at each station. The scientist has eight eggs and is prepared to drop one over each station. What is he testing? He is trying to answer whether or not the egg is more likely to break when dropped over one material as opposed to the other. His hypothesis might have been: The egg will be less likely to break when dropped on linoleum. This is a simple experiment. If the experiment was more complicated, or for example, conducted on a microscopic level, one might want to examine the appropriateness of the instruments utilized and their calibration.

Properly collecting data yields information that appropriately answers the original question. For example, one wouldn't try use a graduated cylinder to measure mass, nor would one use a ruler to measure a microscopic item. Utilizing appropriate measuring devices, using proper units, and careful mathematics will provide strong results. Carefully evaluating and analyzing the data creates a reasonable conclusion. The conclusion needs to be backed up by scientific criteria, then, finally, communicated to the audience.

Skill 1.5 Identify sources of error or uncertainty in an investigation

Unavoidable experimental error is the random error inherent in scientific experiments regardless of the methods used. One source of unavoidable error is measurement and the use of measurement devices. Using measurement devices is an imprecise process because it is often impossible to accurately read measurements. For example, when using a ruler to measure the length of an object, if the length falls between markings on the ruler, we must estimate the true value. Another source of unavoidable error is the randomness of population sampling and the behavior of any random variable. For example, when sampling a population we cannot guarantee that our sample is completely representative of the larger population. In addition, because we cannot constantly monitor the behavior of a random variable, any observations necessarily contain some level of unavoidable error.

Statistical variability is the deviation of an individual in a population from the mean of the population. Variability is inherent in biology because living things are innately unique. For example, the individual weights of humans vary greatly from the mean weight of the population. Thus, when conducting experiments involving the study of living things, we must control for innate variability. Control groups are identical to the experimental group in every way with the exception of the variable being studied. Comparing the experimental group to the control group allows us to determine the effects of the manipulated variable in relation to statistical variability.

COMPETENCY 2.0 UNDERSTAND PROCEDURES FOR GATHERING, RECORDING, ORGANIZING, INTERPRETING, ANALYZING, AND COMMUNICATING SCIENTIFIC DATA AND INFORMATION

Skill 2.1 Use appropriate methods, tools, and technologies for gathering, recording, processing, analyzing, and evaluating data and for communicating the results of scientific investigations

The procedure used to obtain data is important to the outcome. Experiments consist of **controls** and **variables**. A control is the experiment run under normal conditions. The variable includes a factor that is changed. In biology, the variable may be light, temperature, pH, time, etc. The differences in tested variables may be used to make a prediction or form a hypothesis. Only one variable should be tested at a time. One would not alter both the temperature and pH of the experimental subject.

An **independent variable** is one that is changed or manipulated by the researcher. This could be the amount of light given to a plant or the temperature at which bacteria is grown. The **dependent variable** is that which is influenced by the independent variable.

Measurements may be taken in different ways. There is an appropriate measuring device for each aspect of biology. A graduated cylinder is used to measure volume. A balance is used to measure mass. A microscope is used to view microscopic objects. A centrifuge is used to separate two or more parts in a liquid sample. The list goes on, but you get the point. For each variable, there is an appropriate way to measure it. The Internet and teaching guides are virtually unlimited resources for laboratory ideas. You should be imparting on the students the importance of the method with which they conduct the study, the resource they use to do so, the concept of double checking their work, and the use of appropriate units.

Biologists use a variety of tools and technologies to perform tests, collect and display data, and analyze relationships. Examples of commonly used tools include computer-linked probes, spreadsheets, and graphing calculators.

Biologists use computer-linked probes to measure various environmental factors including temperature, dissolved oxygen, pH, ionic concentration, and pressure. The advantage of computer-linked probes, as compared to more traditional observational tools, is that the probes automatically gather data and present it in an accessible format. This property of computer-linked probes eliminates the need for constant human observation and manipulation.

Biologists use spreadsheets to organize, analyze, and display data. For example, conservation ecologists use spreadsheets to model population growth and development, apply sampling techniques and create statistical distributions to analyze relationships. Spreadsheet use simplifies data collection and manipulation and allows the presentation of data in a logical and understandable format.

Graphing calculators are another technology with many applications to biology. For example, biologists use algebraic functions to analyze growth, development and other natural processes. Graphing calculators can manipulate algebraic data and create graphs for analysis and observation. In addition, biologists use the matrix function of graphing calculators to model problems in genetics. The use of graphing calculators simplifies the creation of graphical displays including histograms, scatter plots, and line graphs. Biologists can also transfer data and displays to computers for further analysis. Finally, biologists connect computer-linked probes, used to collect data, to graphing calculators to ease the collection, transmission, and analysis of data.

Skill 2.2 Select appropriate methods and criteria for organizing and displaying data (e.g., tables, graphs, models)

Graphing data
Graphing utilizes numbers to demonstrate patterns. The patterns offer a visual representation, making it easier to draw conclusions.

Graphing is an important skill to visually display collected data for analysis. The two types of graphs most commonly used are the line graph and the bar graph (histogram). Line graphs are set up to show two variables represented by one point on the graph. The X axis is the horizontal axis and represents the independent variable. Independent variables are those that would be present independently of the experiment. A common example of an jndependent variable is time. Time proceeds regardless of anything else that may be occurring. The Y axis is the vertical axis and represents the dependent variable. Dependent variables are manipulated by the experiment, such as the amount of light, or the height of a plant. Graphs should be calibrated at equal intervals. If one space represents one day, the next space may not represent ten days. A "best fit" line is drawn to join the points and may not include all the points in the data. Axes must always be labeled for the graph to be meaningful. A good title will describe both the dependent and the independent variable. Bar graphs are set up similarly in regards to axes, but points are not plotted. Instead, the dependent variable is set up as a bar where the X axis intersects with the Y axis. Each bar is a separate item of data and is not joined by a continuous line.

Skill 2.3 Demonstrate an understanding of the concepts of precision, accuracy, and error with regard to gathering and recording scientific data

Accuracy is the degree of conformity of a measured, calculated quantity to its actual (true) value. Precision, also called reproducibility or repeatability, is the degree to which further measurements or calculations will show the same or similar results.

Accuracy is the degree of veracity while precision is the degree of reproducibility. The best analogy to explain accuracy and precision is the target comparison.

Repeated measurements are compared to arrows that are fired at a target. Accuracy describes the closeness of arrows to the bull's eye at the target center. Arrows that strike closer to the bull's eye are considered more accurate.

All experimental uncertainty is due to either random errors or systematic errors.

Random errors are statistical fluctuations in the measured data due to the precision limitations of the measurement device. Random errors usually result from the experimenter's inability to take the same measurement in exactly the same way to get exactly the same number.

Systematic errors, by contrast, are reproducible inaccuracies that are consistently in the same direction. Systematic errors are often due to a problem which persists throughout the entire experiment.

Systematic and random errors refer to problems associated with making measurements. Mistakes made in the calculations or in reading the instrument are not considered in error analysis.

Skill 2.4 Demonstrate knowledge of the measurement units used in scientific investigations

Science uses the metric system as it is accepted worldwide and allows easier comparison among experiments done by scientists around the world. Learn the following basic units and prefixes:

meter - measure of length
liter - measure of volume
gram - measure of mass

deca-(meter, liter, gram)	= 10X the base unit	deci = 1/10 the base unit
hecto-(meter, liter, gram)	= 100X the base unit	centi= 1/100 the base unit
kilo-(meter, liter, gram)	= 1000X the base unit	milli= 1/1000 the base unit

The common instrument used for measuring volume is the graduated cylinder. The unit of measurement is usually in milliliters (mL). It is important for accurate measure to read the liquid in the cylinder at the bottom of the meniscus, the curved surface of the liquid.

The common instrument used is measuring mass is the triple beam balance. The triple beam balance is measured in as low as tenths of a gram and can be estimated to the hundredths of a gram.

The ruler or meter stick is the most commonly used instrument for measuring length. Measurements in science should always be measured in metric units. Be sure when measuring length that the metric units are used.

Skill 2.5 Identify and evaluate various sources of scientific information (e.g., handbooks, professional journals, popular press, the Internet, community-based resources)

Because people often attempt to use scientific evidence in support of political or personal agendas, the ability to evaluate the credibility of scientific claims is a necessary skill in today's society. In evaluating scientific claims made in the media, public debates, and advertising, one should follow several guidelines.

First, scientific, peer-reviewed journals are the most accepted source for information on scientific experiments and studies. One should carefully scrutinize any claim that does not reference peer-reviewed literature.

Second, the media and those with an agenda to advance (advertisers, debaters, etc.) often overemphasize the certainty and importance of experimental results. One should question any scientific claim that sounds fantastical or overly certain.

Finally, knowledge of experimental design and the scientific method is important in evaluating the credibility of studies. For example, one should look for the inclusion of control groups and the presence of data to support the given conclusions.

COMPETENCY 3.0 UNDERSTAND APPROPRIATE SAFETY PRACTICES, INCLUDING THE SELECTION AND PROPER USE OF MATERIALS, EQUIPMENT, AND TECHNOLOGIES IN SCIENTIFIC INVESTIGATIONS

Skill 3.1 Evaluate equipment, procedures, and settings for potential safety hazards

Safety in the science classroom and laboratory is of paramount importance to the science educator. The following is a general summary of the types of safety equipment that should be made available within a given school system as well as general locations where the protective equipment or devices should be maintained and used. Please note that this is only a partial list and that your school system should be reviewed for unique hazards and site-specific hazards at each facility.The key to maintaining a safe learning environment is through proactive training and regular in-service updates for all staff and students who utilize the science laboratory. Proactive training should include how to **identify potential hazards**, **evaluate potential hazards**, and **how to prevent or respond to hazards**. The following types of training should be considered:

a) Right to Know (OSHA training on the importance and benefits of properly recognizing and safely working with hazardous materials) along with some basic chemical hygiene as well as how to read and understand a material safety data sheet,
b) instruction in how to use a fire extinguisher,
c) instruction in how to use a chemical fume hood,
d) general guidance in when and how to use personal protective equipment (e.g. safety glasses or gloves), and
e) instruction in how to monitor activities for potential impacts on indoor air quality.

It is also important for the instructor to utilize **Material Data Safety Sheets**. Maintain a copy of the material safety data sheet for every item in your chemical inventory. This information will assist you in determining how to store and handle your materials by outlining the health and safety hazards posed by the substance. In most cases the manufacturer will provide recommendations with regard to protective equipment, ventilation and storage practices. This information should be your first guide when considering the use of a new material.

Frequent monitoring and in-service training on all equipment, materials, and procedures will help to ensure a safe and orderly laboratory environment. It will also provide everyone who uses the laboratory the safety fundamentals necessary to discern a safety hazard and to respond appropriately.

Skill 3.2 Maintain safe practices and procedures in all areas related to science instruction

In addition to requirements set forth by your place of employment, the NABT (National Association of Biology Teachers) and ISEF (International Science Education Foundation) have been instrumental in setting parameters for the science classroom. All science labs should contain the following items of **safety equipment** (required by law).

- Fire blanket which is visible and accessible
    - Ground Fault Circuit Interrupters (GFCI) within two feet of water supplies
- Emergency shower capable of providing a continuous flow of water
- Signs designating room exits
- Emergency eye wash station which can be activated by the foot or forearm
- Eye protection for every student and a means of sanitizing equipment
- Emergency exhaust fans providing ventilation to the outside of the building
- Master cut-off switches for gas, electric, and compressed air. Switches must have permanently attached handles. Cut-off switches must be clearly labeled.
- An ABC fire extinguisher
- Storage cabinets for flammable materials

Also recommended, but not required by law:

- Chemical spill control kit
- Fume hood with a motor which is spark proof
- Protective laboratory aprons made of flame retardant material
- Signs which will alert people to potential hazardous conditions
- Containers for broken glassware, flammables, corrosives, and waste.
- Containers should be labeled.

It is the responsibility of teachers to provide a safe environment for their students. Proper supervision greatly reduces the risk of injury and a teacher should never leave a class for any reason without providing alternate supervision. After an accident, two factors are considered; foreseeability and negligence. **Foreseeability** is the anticipation that an event may occur under certain circumstances. **Negligence** is the failure to exercise ordinary or reasonable care. Safety procedures should be a part of the science curriculum and a well managed classroom is important to avoid potential lawsuits

The **"Right to Know Law" statutes** cover science teachers who work with potentially hazardous chemicals. Briefly, the law states that employees must be informed of potentially toxic chemicals. An inventory must be made available if requested. The inventory must contain information about the hazards and properties of the chemicals. Training must be provided in the safe handling and interpretation of the Material Safety Data Sheet.

The following chemicals are potential carcinogens and are not allowed in school facilities:

Acrylonitriel, Arsenic compounds, Asbestos, Bensidine, Benzene, Cadmium compounds, Chloroform, Chromium compounds, Ethylene oxide, Ortho-toluidine, Nickel powder, Mercury.

Skill 3.3 Respond to accidents by applying first-response procedures, including first aid

All students and staff should be trained in first aid in the science classroom and laboratory. Please remember to always report all accidents, however minor, to the lab instructor immediately. In most situations 911 should immediately be called. Please refer to your school's specific safety plan for accidents in the classroom and laboratory. The classroom/laboratory should have a complete first-aid kit with supplies that are up-to-date and checked frequently for expiration.

Know the location and use of fire extinguishers, eye-wash stations, and safety showers in the lab.

Do not attempt to smother a fire in a beaker or flask with a fire extinguisher. The force of the stream of material from it will turn over the vessel and result in a bigger fire. Just place a watch glass or a wet towel over the container to cut off the supply of oxygen.

If your clothing is on fire, **do not run** because this only increases the burning. It is normally best to fall on the floor and roll over to smother the fire. If a student, whose clothing is on fire panics and begins to run, attempt to get the student on the floor and roll over to smother the flame. If necessary, use the fire blanket or safety shower in the lab to smother the fire.

Students with long hair should put their hair in a bun or a ponytail to prevent their hair from catching on fire.

Below are common accidents that everyone who uses the laboratory should be trained in how to respond.

Burns (Chemical or Fire) – Use deluge shower for 15 minutes.

Burns (Clothing on fire) – Use safety shower immediately. Keep victim immersed 15 minutes to wash away both heat and chemicals. All burns should be examined by medical personnel.

Chemical spills – Chemical spills on hands or arms should be washed immediately with soap and water. Washing hands should become an instinctive response to any chemical spilled on hands. Spills that cover clothing and other parts of the body should be drenched under the safety shower. If strong acids or bases are spilled on clothing, the clothing should be removed. If a large area is affected, remove clothing and immerse victim in the safety shower. If a small area is affected, remove article of clothing and use deluge shower for 15 minutes.

Eyes (chemical contamination) – Hold the eye wide open and flush with water from the eye wash for about 15 minutes. Seek medical attention.

Ingestion of chemicals or poisoning – See antidote chart on wall of lab for general first-aid directions. The victim should drink large amounts of water. All chemical poisonings should receive medical attention.

Skill 3.4 Identify various sources of information about safety, legal issues, and the proper use, storage, and disposal of scientific materials (e.g., state and federal regulations and guidelines, material safety data sheets)

Storing, identifying, and disposing of chemicals and biological materials

All laboratory solutions should be prepared as directed in the lab manual. Care should be taken to avoid contamination. All glassware should be rinsed thoroughly with distilled water before using and cleaned well after use. All solutions should be made with distilled water as tap water contains dissolved particles that may affect the results of an experiment. Unused solutions should be disposed of according to local disposal procedures.

The "Right to Know Law" covers science teachers who work with potentially hazardous chemicals. Briefly, the law states that employees must be informed of potentially toxic chemicals. An inventory must be made available if requested. The inventory must contain information about the hazards and properties of the chemicals. This inventory is to be checked against the "Substance List". Training must be provided on the safe handling and interpretation of the Material Safety Data Sheet.

The following chemicals are potential carcinogens and not allowed in school facilities: Acrylonitriel, Arsenic compounds, Asbestos, Bensidine, Benzene, Cadmium compounds, Chloroform, Chromium compounds, Ethylene oxide, Ortho-toluidine, Nickle powder, and Mercury.

Chemicals should not be stored on bench tops or near heat sources. They should be stored in groups based on their reactivity with one another and in protective storage cabinets. All containers within the lab must be labeled. Suspect and known carcinogens must be labeled as such and segregated within trays to contain leaks and spills.

Chemical waste should be disposed of in properly labeled containers. Waste should be separated based on their reactivity with other chemicals. Biological material should never be stored near food or water used for human consumption. All biological material should be appropriately labeled. All blood and body fluids should be put in a well-contained container with a secure lid to prevent leaking. All biological waste should be disposed of in biological hazardous waste bags.

In addition to the safety laws set forth by the government for equipment necessary to the lab, OSHA (Occupational Safety and Health Administration) has helped to make environments safer by instituting signs that are bilingual. These signs use pictures in addition to words and feature eye-catching colors. Some of the best-known examples are exit, restrooms, and handicap accessible signs.

Of particular importance to laboratories are diamond safety signs, prohibitive signs, and triangle danger signs. Each sign encloses a descriptive picture.

As a teacher, you should utilize a MSDS (Material Safety Data Sheet) whenever you are preparing an experiment. It is designed to provide people with the proper procedures for handling or working with a particular substance. MSDS's include information such as physical data (melting point, boiling point, etc.), toxicity, health effects, first aid, reactivity, storage, disposal, protective gear, and spill/leak procedures. These are particularly important if a spill or other accident occurs. You should review a few, available commonly online, and understand the listing procedures. Material safety data sheets are available directly from the company of acquisition or the internet. The manuals for equipment used in the lab should be read and understood before using the equipment.

Skill 3.5 Demonstrate knowledge of the safe and proper use of science tools, equipment, chemicals, materials, and technology in scientific inquiry (e.g., computers, scientific instrumentation)

Light microscopes are commonly used in high school laboratory experiments. Total magnification is determined by multiplying the ocular (usually 10X) and the objective (usually 10X on low, 40X on high) lenses. Several procedures should be followed to properly care for this equipment.

-Clean all lenses with lens paper only.
-Carry microscopes with two hands; one on the arm and one on the base.
-Always begin focusing on low power, then switch to high power.
-Store microscopes with the low power objective down.
-Always use a coverslip when viewing wet mount slides.
-Bring the objective down to its lowest position then focus moving up to avoid breaking the slide or scratching the lens.

Wet mount slides should be made by placing a drop of water on the specimen and then putting a glass coverslip on top of the drop of water. Dropping the coverslip at a forty-five degree angle will help in avoiding air bubbles.

Chromatography uses the principles of capillarity to separate substances such as plant pigments. Molecules of a larger size will move slower up the paper, whereas smaller molecules will move more quickly producing lines of pigment.

An **indicator** is any substance used to assist in the classification of another substance. An example of an indicator is litmus paper. Litmus paper is a way to measure whether a substance is acidic or basic. Blue litmus turns pink when an acid is placed on it and pink litmus turns blue when a base is placed on it. pH paper is a more accurate measure of pH, with the paper turning different colors depending on the pH value.

Spectrophotometry measures percent of light at different wavelengths absorbed and transmitted by a pigment solution.

Centrifugation involves spinning substances at a high speed. The more dense part of a solution will settle to the bottom of the test tube, where the lighter material will stay on top. Centrifugation is used to separate blood into blood cells and plasma, with the heavier blood cells settling to the bottom.

Electrophoresis uses electrical charges of molecules to separate them according to their size. The molecules, such as DNA or proteins are pulled through a gel towards either the positive end of the gel box (if the material has a negative charge) or the negative end of the gel box (if the material has a positive charge). DNA is negatively charged and moves towards the positive charge.

Skill 3.6 Apply various procedures for using and caring for living organisms in an ethical and humane manner according to the standards of the National Association of Biology Teachers

Live specimens - No dissections may be performed on living mammalian vertebrates or birds. Lower order life and invertebrates may be used. Biological experiments may be done with all animals except mammalian vertebrates or birds. No physiological harm may result to the animal. All animals housed and cared for in the school must be handled in a safe and humane manner. Animals are not to remain on school premises during extended vacations unless adequate care is provided. Many state laws stipulate that any instructor who intentionally refuses to comply with the laws may be suspended or dismissed. Pathogenic organisms must never be used for experimentation. Students should adhere to the following rules at all times when working with microorganisms to avoid accidental contamination:

1. Treat all microorganisms as if they were pathogenic.
2. Maintain sterile conditions at all times

COMPETENCY 4.0 UNDERSTAND SHARED CONCEPTS, THEMES, AND METHODS AMONG SCIENTIFIC AND OTHER DISCIPLINES AND THE INTERDEPENDENCE OF SCIENCE AND TECHNOLOGY

Skill 4.1 Identify common scientific concepts and themes (e.g., change, systems, models, form and function) that link and unify science fields

The following are the concepts and processes generally recognized as common to all scientific disciplines:

Systems, order, and organization
Evidence, models, and explanation
Constancy, change, and measurement
Evolution and equilibrium
Form and function

Because the natural world is so complex, the study of science involves the **organization** of items into smaller groups based on interaction or interdependence. These groups are called **systems**. Examples of organization are the periodic table of elements and the six-kingdom classification scheme for living organisms. Examples of systems are the solar system, cardiovascular system, Newton's laws of force and motion, and the laws of conservation.

Order refers to the behavior and measurability of organisms and events in nature. The arrangement of planets in the solar system and the life cycle of bacterial cells are examples of order.

Scientists use **evidence** and **models** to form **explanations** of natural events. Models are miniaturized representations of a larger event or system. Evidence is anything that furnishes proof.

Constancy and **change** describe the observable properties of natural organisms and events. Scientists use different systems of **measurement** to observe change and constancy. For example, the freezing and melting points of given substances and the speed of sound are constant under constant conditions. Growth, decay, and erosion are all examples of natural change.

Evolution is the process of change over a long period of time. While biological evolution is the most common example, one can also classify technological advancement, changes in the universe, and changes in the environment as evolution.

Equilibrium is the state of balance between opposing forces of change. Homeostasis and ecological balance are examples of equilibrium.

Form and **function** are properties of organisms and systems that are closely related. The function of an object usually dictates its form and the form of an object usually facilitates its function. For example, the form of the heart (e.g. muscle, valves) allows it to perform its function of circulating blood through the body.

Skill 4.2 Demonstrate knowledge of the general characteristics and properties of systems and how system components and different systems interact (e.g., feedback)

Groups of related organs are organ systems. Organ systems consist of organs working together to perform a common function. The commonly recognized organ systems of animals include the reproductive system, nervous system, circulatory system, respiratory system, lymphatic system (immune system), endocrine system, excretory system, muscular system, digestive system, integumentary system, and skeletal system. In addition, organ systems are interconnected and a single system rarely works alone to complete a task.

One obvious example of the interconnectedness of organ systems is the relationship between the circulatory and respiratory systems. As blood circulates through the organs of the circulatory systems, it is re-oxygenated in the lungs of the respiratory system. Another example is the influence of the endocrine system on other organ systems. Hormones released by the endocrine system greatly influence processes of many organ systems including the nervous and reproductive systems.

In addition, bodily response to infection is a coordinated effort of the lymphatic (immune system) and circulatory systems. The lymphatic system produces specialized immune cells, filters out disease-causing organisms, and removes fluid waste from in and around tissue. The lymphatic system utilizes capillary structures of the circulatory system and interacts with blood cells in a coordinated response to infection.

The pituitary gland and hypothalamus respond to varying levels of hormones by increasing or decreasing production and secretion. High levels of a hormone decrease the production and secretion pathways, while low levels of a hormone increase the production and secretion pathways.

"Fight or flight" refers to the human body's response to stress or danger. Briefly, as a response to an environmental stressor, the hypothalamus releases a hormone that acts on the pituitary gland, triggering the release of another hormone, adrenocorticotropin (ACTH), into the bloodstream. ACTH then signals the adrenal glands to release the hormones cortisol, epinephrine, and norepinephrine. These three hormones act to ready the body to respond to a threat by increasing blood pressure and heart rate, speeding reaction time, diverting blood to the muscles, and releasing glucose for use by the muscles and brain. The stress response hormones also down-regulate growth, development, and other nonessential functions. Cortisol completes the "fight or flight" feedback loop by acting on the hypothalamus to stop hormonal production after the threat has passed.

Finally, the muscular and skeletal systems are closely related. Skeletal muscles attach to the bones of the skeleton and drive movement of the body.

Skill 4.3 Demonstrate an understanding of how models are used in science and methods for evaluating the strengths and weaknesses of scientific models

The model is a basic element of the scientific method. Many things in science are studied with models. A model is any simplification or substitute for what we are actually studying, understanding or predicting. A model is a substitute, but it is similar to what it represents. We encounter models at every step of our daily living. The Periodic Table of the elements is a model chemists use for predicting the properties of the elements. Physicists use Newton's laws to predict how objects will interact, such as planets and spaceships. In geology, the continental drift model predicts the past positions of continents. Sample, ideas, and methods are all examples of models. At every step of scientific study models are extensively use. The primary activity of the hundreds of thousands of US scientists is to produce new models, resulting in tens of thousands of scientific papers published per year.

Types of models:

* Scale models: some models are basically downsized or enlarged copies of their target systems like the models of protein or DNA.

* idealized models: An idealization is a deliberate simplification of something complicated with the objective of making it easier to understand. Some examples are frictionless planes, point masses or isolated systems.

*Analogical models: standard examples of analogical models are the billiard model of a gas, the computer model of the mind, or the liquid drop model of the nucleus.

*Phenomenological models: These are usually defined as models that are independent of theories.

*Data models: These are corrected, rectified, regimented, and in many instances, idealized version of the data we gain from immediate observation (raw data).

*Theory models: Any structure is a model if it represents an idea (theory). An example of this is a flow chart, which summarizes a set of ideas.

Uses of models:

1. Models are crucial for understanding the structure and function of processes in science.
2. Models help us to visualize the organs/systems they represent just like putting a face to person.
3. Models are very useful to predict and foresee future events like hurricanes

Limitations:

1. Although models are very useful to us, they can never replace the real thing.
2. Models are not exactly like the real item they represent.
3. Caution must be exercised before presenting the models to the class, as they may not be accurate.
4. It is the responsibility of the educator to analyze the model critically for the proportions, content value, and other important data.
5. One must be careful about the representation style. This style differs from person to person.

Skill 4.4 Analyze the interrelationships among science and other disciplines (e.g., mathematics, arts, social studies, language arts)

Math, science, and technology have common themes in how they are applied and understood. All three use models, diagrams, and graphs to simplify a concept for analysis and interpretation. Patterns observed in these systems lead to predictions based on these observations. Another common theme among these three systems is equilibrium. Equilibrium is a state in which forces are balanced, resulting in stability. Static equilibrium is stability due to a lack of changes and dynamic equilibrium is stability due to a balance between opposite forces.

The fundamental relationship between the natural and social sciences is the use of the scientific method and the rigorous standards of proof that both disciplines require. This emphasis on organization and evidence separates the sciences from the arts and humanities. Natural science, particularly biology, is closely related to social science, the study of human behavior. Biological and environmental factors often dictate human behavior and accurate assessment of behavior requires a sound understanding of biological factors.

Skill 4.5 Identify concepts and methods that are common to science and technology and analyze the interdependence of science and technology

Biological science is closely connected to technology and the other sciences and greatly impacts society and everyday life. Scientific discoveries often lead to technological advances and, conversely, technology is often necessary for scientific investigation and advances in technology often expand the reach of scientific discoveries. In addition, biology and the other scientific disciplines share several unifying concepts and processes that help unify the study of science. Finally, because biology is the science of living systems, biology directly impacts society and everyday life.

Science and technology, while distinct concepts, are closely related. Science attempts to investigate and explain the natural world, while technology attempts to solve human adaptation problems. Technology often results from the application of scientific discoveries, and advances in technology can increase the impact of scientific discoveries. For example, Watson and Crick used science to discover the structure of DNA and their discovery led to many biotechnological advances in the manipulation of DNA. These technological advances greatly influenced the medical and pharmaceutical fields. The success of Watson and Crick's experiments, however, was dependent on the technology available. Without the necessary technology, the experiments would have failed.

The combination of biology and technology has improved the human standard of living in many ways. However, the negative impact of increasing human life expectancy and population on the environment is problematic. In addition, advances in biotechnology (e.g. genetic engineering, cloning) produce ethical dilemmas that society must consider. Biologists use a variety of tools and technologies to perform tests, collect and display data and analyze relationships. Examples of commonly used tools include computer-linked probes, spreadsheets, and graphing calculators.

Biologists use computer-linked probes to measure various environmental factors including temperature, dissolved oxygen, pH, ionic concentration, and pressure. The advantage of computer-linked probes, as compared to more traditional observational tools, is that the probes automatically gather data and present it in an accessible format. This property of computer-linked probes eliminates the need for constant human observation and manipulation.

Biologists use spreadsheets to organize, analyze, and display data. For example, conservation ecologists use spreadsheets to model population growth and development, apply sampling techniques, and create statistical distributions to analyze relationships. Spreadsheet use simplifies data collection and manipulation and allows the presentation of data in a logical and understandable format.

Graphing calculators are another technology with many applications to biology. For example, biologists use algebraic functions to analyze growth, development and other natural processes. Graphing calculators can manipulate algebraic data and create graphs for analysis and observation. In addition, biologists use the matrix function of graphing calculators to model problems in genetics. The use of graphing calculators simplifies the creation of graphical displays including histograms, scatter plots and line graphs. Biologists can also transfer data and displays to computers for further analysis. Finally, biologists connect computer-linked probes, used to collect data, to graphing calculators to ease the collection, transmission and analysis of data.

COMPETENCE 5.0 UNDERSTAND THE HISTORY OF SCIENCE AND THE INTERRELATIONSHIPS OF SCIENCE AND SOCIETY

Skill 5.1 Identify key events in the history of science and the science contributions of people from a variety of social and ethnic backgrounds

The history of biology follows man's understanding of the living world from the earliest recorded history to modern times. Though the concept of biology as a field of science arose only in the 19th century, its origins could be traced back to the ancient Greeks (Galen and Aristotle).

During the Renaissance and Age of Discovery, renewed interest in the rapidly increasing number of known organisms generated a lot of interest in biology.

Andreas Vesalius (1514-1564) was a Belgian anatomist and physician whose dissections of the human body and the descriptions of his findings helped to correct the misconceptions of science. The books Vesalius wrote on anatomy were the most accurate and comprehensive anatomical texts of time.

Anton van Leeuwenhoek is known as the father of microscopy. In the 1650s, Leeuwenhoek began making tiny lenses that gave magnifications up to 300x. He was the first to see and describe bacteria, yeast, plants, and the microscopic life found in water. Over the years, light microscopes have advanced to produce greater clarity and magnification. The scanning electron microscope (SEM) was developed in the 1950's. Instead of light, a beam of electrons passes through the specimen. Scanning electron microscopes have a resolution about one thousand times greater than light microscopes. The disadvantage of the SEM is that the chemical and physical methods used to prepare the sample result in the death of the specimen.

Carl Von Linnaeus (1707-1778), a Swedish botanist, physician, and zoologist, is well known for his contributions in ecology and taxonomy. Linnaeus is famous for his binomial system of nomenclature in which each living organism has two names, a genus and a species name. He is considered the father of modern ecology and taxonomy.

In the late 1800's, Pasteur discovered the role of microorganisms in the cause of disease, pasteurization, and the rabies vaccine. Koch took his observations one step further by postulating that specific diseases were caused by specific pathogens. **Koch's postulates** are still used as guidelines in the field of microbiology. They state that the same pathogen must be found in every diseased person, the pathogen must be isolated and grown in culture, the disease must be induced in experimental animals from the culture, and the same pathogen must be isolated from the experimental animal.

In the 18th century, many fields of science like botany, zoology and geology began to evolve as scientific disciplines in the modern sense.

In the 20th century, the rediscovery of Mendel's work led to the rapid development of genetics by Thomas Hunt Morgan and his students.

DNA structure was another key event in biological study. In the 1950's, James Watson and Francis Crick discovered the structure of a DNA molecule as that of a double helix. This structure made it possible to explain DNA's ability to replicate and to control the synthesis of proteins.

Following the cracking of the genetic code, biology has largely split between organismal biology-consisting of ecology, ethology, systematics, paleontology, evolutionary biology, developmental biology, and other disciplines that deal with whole organisms or group of organisms, and the disciplines related to molecular biology, which include cell biology, biophysics, biochemistry, neuroscience, and immunology.

The use of animals in biological research has expedited many scientific discoveries. Animal research has allowed scientists to learn more animal biological systems, including the circulatory and reproductive systems. One significant use of animals is for the testing of drugs, vaccines, and other products (such as perfumes and shampoos) before use or consumption by humans. There are both significant pros and cons of animal research. The debate about the ethical treatment of animals has been ongoing since the introduction of animals to research. Many people believe the use of animals in research is cruel and unnecessary. Animal use is federally and locally regulated. The purpose of the Institutional Animal Care and Use Committee (IACUC) is to oversee and evaluate all aspects of an institution's animal care and use program.

Skill 5.2 Demonstrate an understanding of the influence of social and cultural factors on science and technology

Curiosity is the heart of science. Maybe this is why so many diverse people are drawn to it. In the area of zoology one of the most recognized scientists is Jane Goodall. Miss Goodall is known for her research with chimpanzees in Africa. Jane has spent many years abroad conducting long term studies of chimp interactions, and returns from Africa to lecture and provide information about Africa, the chimpanzees, and her institute located in Tanzania.

In the area of chemistry we recognize Dorothy Crowfoot Hodgkin. She studied at Oxford and won the Nobel Prize of Chemistry in 1964 for recognizing the shape of vitamin B12.

Have you ever heard of Florence Nightingale? She was a true person living in the 1800's and she shaped the nursing profession. Florence was born into wealth and shocked her family by choosing to study health reforms for the poor in lieu of attending the expected social events. Florence studied nursing in Paris and became involved in the Crimean war. The British lacked supplies and the secretary of war asked for Florence's assistance. She earned her nickname walking the floors at night checking on patients and writing letters to British officials demanding supplies.

In 1903 the Nobel Prize in Physics was jointly awarded to three individuals: Marie Curie, Pierre Curie, and Becquerel. Marie was the first woman ever to receive this prestigious award. In addition, she received the Nobel Prize in chemistry in 1911, making her the only person to receive two Nobel awards in science. Ironically, her cause of death in 1934 was of overexposure to radioactivity, the research for which she was so respected.

Neil Armstrong is an American icon. He will always be symbolically linked to our aeronautics program. This astronaut and naval aviator is known for being the first human to set foot on the Moon.

Sir Alexander Fleming was a pharmacologist from Scotland who isolated the antibiotic penicillin from a fungus in 1928. Fleming also noted that bacteria developed resistance whenever too little penicillin was used or when it was used for too short a period, a key problem we still face today.

It is important to realize that many of the most complex scientific questions have been answered in a collaborative form. The human genome project is a great example of research conducted and shared by multiple countries worldwide.

It is also interesting to note that because of differing cultural beliefs, some cultures may be more likely to allow areas of research that other cultures may be unlikely to examine.

Society and culture are very closely intertwined. Together they have influenced every aspect of the human life. Science and technology are no exception to this.

Let us examine the influence of social and cultural factors on science first and then we will see their effect on technology.

The influence of social and cultural factors on science is profound. In a way we can say that society has changed the face of science by absorbing scientific innovations. Science has always been a big part of society. The difference is that in ancient societies, people did not realize that it was science, but took it as a part of their lives. In the modern society, everything has a label and a name, so that people are aware of science and other disciplines.

Societies have had trouble accepting science, especially where the science exposed some cultural aspects as myths. There was a big dilemma as to accept the proven facts provided by scientific investigations or cling to cultural norms. This went on for centuries. It took a long time for societies to accept these facts and to leave some of the cultural practices behind or to modify them. At the same time, we must give full credit to cultural practices, which are scientifically correct, but are sometimes connected to religion and taken very seriously by believers. We can conclude that there are two factors - one is cultural practices by societies which are scientifically correct and the second one is cultural practices which have no scientific foundation (myths and superstitions). A society's progress depends on distinguishing between these two. Some indigenous societies suffered when they were not quick adjust, since their cultures are very ancient and the people found it difficult to accept new challenges and adapt to new changes. At the same time, ancient cultures like the Chinese, Egyptian, Greek, Asian, and Indian had well developed science that was recorded in their ancient writings.

Let's take a look at the effect of society and culture on technology. If we compare science to a volcano, technology is like lava spewing out of the volcano. This was the scenario in the last few centuries in terms of rapid strides in the development of technology. Technology greatly influenced society and culture and at the same time, science and culture exercised their influence on technology. It is like a two-way street.

It became extremely difficult for some societies to come to terms with technological advances. Even today, some cultures are not using modern technology, but at the same time, they are using technology in principle, such as using simple machines for farming rather than using complex machines like tractors. Other cultures have so readily adapted to technology that lives are intertwined with it; intertwined so much that we are utilize the computer, television, microwave, dishwasher, washing machine and cell phone on a daily basis. It is surprising to realize that we began with no technology and now are enslaved to it. Cultures that are not in tune with modern technology are falling behind. It is often argued that to live without technology yields peace of mind, serenity and happiness, but they are also losing valuable opportunities in this age of communication.

Positive contributions of technology are that it revolutionized education, medicine, communication and travel. The world seems to have shrunk in that we don't seem as far apart and we have means to stay connected. As a result of this technology, man is exploring space, to find out what it is like and to learn and gain knowledge, which used to be elusive and as distant as the planets themselves.

When we take a critical look at these facts, we have to commend societies for trying to keep their own culture, as culture is a very important aspect of humanity.

We also need to appreciate cultures that accepted or incorporated science and technology.

Skill 5.3 Demonstrate the ability to distinguish between the ethical and unethical uses of science (e.g., the use of proper protocol)

To understand scientific ethics, we need to have a clear understanding of ethics. Ethics is defined as a system of public, general rules for guiding human conduct (Gert, 1988). The rules are general because they are supposed to all people at all times and they are public because they are not secret codes or practices.

Philosophers have given a number of moral theories to justify moral rules, which range from utilitarianism, a theory of ethics that prescribes the quantitative maximization of good consequences for a population. It is a form of consequentialism. This theory was proposed by Mozi, a Chinese philosopher who lived during BC 471-381. Kantianism, a theory proposed by Immanuel Kant, a German philosopher who lived during 1724-1804, ascribes intrinsic value to rational beings and is the philosophical foundation of contemporary human rights to social contract theory, a view of the ancient Greeks which states that the person's moral and or political obligations are dependent upon a contract or agreement between them to form society.

The following are some of the guiding principles of scientific ethics:

1. Scientific Honesty: not to fraud, fabricate or misinterpret data for personal gain
2. Caution: to avoid errors and sloppiness in all scientific experimentation
3. Credit: give credit where credit is due and not to copy
4. Responsibility: only to report reliable information to public and not to mislead in the name of science
5. Freedom: freedom to criticize old ideas, question new research and conduct research.

Many more principles could be added to this list. Though these principles seem straightforward and clear it is very difficult to put them into practice since they could be interpreted in more ways than one. Nevertheless, it is not an excuse for scientists to overlook these guiding principles of scientific ethics.

Scientists are expected to show good conduct in their scientific pursuits. Conduct here refers to all aspects of scientific activity including experimentation, testing, education, data evaluation, data analysis, data storing, peer review, government funding and staffing.

The common ethical code described above could be applied to many areas including science. When the general code is applied to a particular area of human life, it then becomes an institutional code. Hence, scientific ethics is an institutional code of conduct that reflects the chief concerns and goals of science.

To discuss scientific ethics, we can look at natural phenomena like rain. Rain in the normal sense is extremely useful to us and it is absolutely important that there is a water cycle. When rain gets polluted with car exhaust, it becomes acid rain. Here lies the ethical issue of releasing all these pollutants into the atmosphere. Should the scientists communicate the whole truth about acid rain or withhold some information because it may alarm the public. There are many issues like this. Whatever may be the case; scientists are expected to be honest and forthright with the public.

Skill 5.4 Identify the effects of scientific and technological developments on the environment, human biology, society, and culture

In the last century, the advances in the fields of science and technology were amazing and have changed the lives of human beings forever. Lifestyles were greatly affected and society experienced dramatic changes. People started to take science technology very seriously. The advances in these two interrelated fields are no longer the domain of the elite and sophisticated. The average person started to use the advances in the field of technology in their daily lives. Because of this, the societal structure is changing rapidly to the extent that even young children are using technology.

With any rapid change, there are always good and bad things associated with it. Caution and care are the two words we need to associate with these giant strides in technology. At the same time, we need high technology in our lives and we can't afford to not make use of these developments and reap the benefits for the good of humanity.

Our environment in which we live, human biology, society at large, and our culture are being affected. Let us take each point and examine very carefully the effects of science and technology on the above.

1. Environment:

The environment we live in is constantly and rapidly undergoing tremendous changes.

The positive effects include the ability to predict hurricanes, measuring the changes in terms of radioactivity present in our environment, the remedial measures for that problem, predicting the levels of gases like carbon monoxide, carbon dioxide, and other harmful gases, various estimates like the green house effect, ozone layer, UV radiation, to name a few. With the help of modern technology, it is possible to know their quantities and to monitor and plan and implement measures to deal with them. Even with the most advanced technology available to us, it is impossible to go back to the clean, green earth, since man has made mark on it in a negative way. It is possible to a limited extent to alleviate the problem, but it is impossible to eradicate it.

The negative aspects of the effect of technology on our environment are numerous. The first and foremost is pollution of various kinds such as water, air and noise. The greenhouse effect, the indiscriminate use of fertilizers, the spraying of pesticides, the use of various additives to our food, deforestation, and the unprecedented exploitation of non renewable energy resources. As we discussed earlier, it is not possible to solve these problems with money or human resources, but educating the society and making them aware of these negative aspects will go a long way. For example, as teachers we need to educate students about using natural resources cautiously, trying to save those resources, and we also need to teach that little steps in the right direction will go a long way, such as car pooling, not wasting paper, whenever possible, to walk (if it is safe). It is important to teach the students to value trees and plants as they recycle oxygen back into the atmosphere.

2. Human biology:

The strides made by science and technology have lasting effects on human biology. A few examples are organ transplants, in vitro fertilization, cloning, new drugs, new understanding of various diseases using scientific knowledge, cosmetic surgery, reconstructive surgery, use of computers in operations, lasers in medicine and forensic science. These changes have made lasting difference to humanity.

As always there are pros and cons to these changes.

The positive aspects are that people with organ transplants have renewed hope. Their lifespans are increased and their quality of life has changed with the use of technology such as pacemakers. Couples who experienced infertility are having babies now. Corrective and cosmetic surgery are giving new confidence to patients. Glasses to correct vision problems are being replaced slowly by laser surgery.

The negative aspects are paternity issues arising out of in vitro fertilization, some medical blunders, which are expensive and heart wrenching (when a wrong egg is implanted), the indiscriminate use of corrective and cosmetic surgery and older mothers who may die and leave and young orphans.

3. Society:

Society is not the same as it used to be even 25 years ago. The use of technology has changed our patterns of lifestyle, our behavior, our ethical and moral thinking, our economy and career opportunities, to name a few.

The positive effects are the booming economy due to the high tech industry, more career opportunities for people to select, raising of the standard of living, prolonged life with quality, closeness even though we are separated by thousands of kilometers/miles and quicker and faster communication. The computer has contributed a lot to these changes. Normal household chores are being done by machines, which is a cost effective and time saving means for upkeep of kitchen and home, giving relief to a busy lifestyle.

The negative aspects are far reaching. The breakdown in family structure could be attributed partly to high tech. Family meals and family togetherness are being replaced with gadgets. Some would argue that as a result of this our young people are becoming insecure, indirectly affecting their problem solving skills. Young people are becoming increasingly vulnerable due to Internet programs including chat rooms and online pornography. There must be stringent measures to protect our younger generation from these Internet predators. The effects of various high tech gadgets are not entirely positive. Constant game playing utilizing new technology such as Gameboy and Xbox, encourage a sedentary lifestyle and childhood obesity.

4. Culture:

This is a very sensitive yet very important issue. The above listed factors are affecting the culture of people.

The positive aspects are that technology is uniting us to a certain extent. For exampe it is possible to communicate with a person of any culture without seeing them face to face. It makes business and personal communication much easier over long distances. Some people were not comfortable with communicating with other cultures, since they were closed societies, but e-mail has changed that. When we all use the same pieces of technology, we relate better and a common ground is established. The Internet can definitely boast of some successful cross-cultural marriages. Sharing opinions and information has also been enhanced.

With modern technology, travel is changing the way we think and increasing career opportunities. It is helping us to understand other cultures, to find different ways of doing the same thing and to learn the positive values of other cultures.

The negative aspects include moral and ethical values, as increased awareness is allowing for a new wave of thinking. Care must be exercised as to how much of our past culture we are willing to trade for the modern trends. Positive aspects of any culture must be guarded carefully and passed on to generations to come.

On the whole, we can safely conclude that science and technology are part of our lives and we must always exercise caution and be careful when we are adapting to new ideas and new thinking. It is possible that awareness and incorporation of other cultural practices will make us a better nation, which our founding fathers envisioned and dreamed of.

Skill 5.5 Identify careers in science and evaluate reasons why people choose science as a career

Science is an interesting, innovative, and thoroughly enjoyable subject. Science careers are challenging and stimulating. The possibilities for scientific careers are endless. Currently, the sky is the limit for opportunities in science for anybody who is interested in that kind of challenge.

Why do people choose careers in science? This is a very important question. The reasons are manifold.

1. A passion for science
2. A desire to experiment and gain knowledge and/ or contribute to society's betterment
3. An inquiring mind
4. Wanting to work in a team

There are a number of opportunities in science. For the sake of ease and convenience, they are grouped under various categories as follows:

1. Biological sciences
2. Physical sciences
3. Earth science
4. Space science
5. Forensic science
6. Medical science
7. Agricultural science

Let's take each category and examine the opportunities available.

1. **Biological Science d**eals with the study of living organisms and their life cycles. People who are interested in studying living things opt for these kinds of careers.
   * Botanist - studies plants, for someone who is interested in plants
   * Microbiologist: studies microscopic organisms, their uses, harmful effects and diseases

2. **Physical Science** includes careers dealing with various branches of Physical Science (the study of matter and energy)
   * Analytical Chemist
   * Biochemist
   * Chemist
   * Physicist

3. **Earth Science** is the study of the earth, its changes over the years and natural disasters such as earthquakes and hurricanes.
   * Geologist
   * Meteorologist
   * Oceanographer
   * Seismologist
   * Volcanologist

4. **Space Science** is the study of space, the universe and the planets. Somebody who is very strong in Math and Physics and who wants to know about space go for these opportunities.
   * Astrophysicist
   * Space Scientist

5. **Forensic Science** involves solving crimes using various techniques
   * Forensic Pathologist

6. **Medical Science:**
   * Biomedical science
   * Clinical Scientist

7. **Agricultural Science**
   * Agriculturist: grows crops using modern methods
   * Agricultural Service Industry: the business side of agriculture including marketing, etc.
   * Agronomist: studies soil, also known as Soil Scientist
   * Veterinary science: deals with animals

There are so many career opportunities available to our youngsters, but it is up to them to choose the right career. The students need to be made aware of the connection between today's learning and their future life. It is especially important to impress upon them how science is everywhere, and its truly useful applications. When this is made clear to them, they will more seriously consider science as a career.

SUBAREA II. LIFE SCIENCE

COMPETENCY 6.0 UNDERSTAND BASIC CONCEPTS OF CELL BIOLOGY

Skill 6.1 Identify characteristics and functions of biologically important compounds (e.g., carbohydrates, lipids, proteins, nucleic acids)

A compound consists of two or more elements. There are four major chemical compounds found in the cells and bodies of living things. These include carbohydrates, lipids, proteins and nucleic acids.

Monomers are the simplest unit of structure. **Monomers** can be combined to form **polymers**, or long chains, making a large variety of molecules possible. Monomers combine through the process of condensation reaction (also called dehydration synthesis). In this process, one molecule of water is removed between each of the adjoining molecules. In order to break the molecules apart in a polymer, water molecules are added between monomers, thus breaking the bonds between them. This is called hydrolysis.

Carbohydrates contain a ratio of two hydrogen atoms for each carbon and oxygen $(CH_2O)_n$. Carbohydrates include sugars and starches. They function in the release of energy. **Monosaccharides** are the simplest sugars and include glucose, fructose, and galactose. They are major nutrients for cells. In cellular respiration, the cells extract the energy in glucose molecules. **Disaccharides** are made by joining two monosaccharides by condensation to form a glycosidic linkage (covalent bond between two monosaccharides). Maltose is formed from the combination of two glucose molecules, lactose is formed from joining glucose and galactose, and sucrose is formed from the combination of glucose and fructose. **Polysaccharides** consist of many joined monomers. They are used as storage material, hydrolyzed as needed to provide sugar for cells or building material for structures protecting the cell. Examples of polysaccharides include starch, glycogen, cellulose and chitin.

Starch - major energy storage molecule in plants. It is a polymer consisting of glucose monomers.
Glycogen - major energy storage molecule in animals. It is made up of many glucose molecules.
Cellulose - found in plant cell walls, its function is structural. Many animals lack the enzymes necessary to hydrolyze cellulose, so it simply adds bulk (fiber) to the diet.
Chitin - found in the exoskeleton of arthropods and fungi. Chitin contains an amino sugar (glycoprotein).

Lipids are composed of glycerol (an alcohol) and three fatty acids. Lipids are **hydrophobic** (water fearing) and will not mix with water. There are three important families of lipids, fats, phospholipids and steroids.

Fats consist of glycerol (alcohol) and three fatty acids. Fatty acids contain long carbon skeletons. The nonpolar carbon-hydrogen bonds in the tails of fatty acids are why they are hydrophobic. Fats are solids at room temperature and come from animal sources (butter, lard).

Phospholipids are a vital component in cell membranes. In a phospholipid, one or two fatty acids are replaced by a phosphate group linked to a nitrogen group. They consist of a **polar** (charged) head that is hydrophilic or water loving and a **nonpolar** (uncharged) tail which is hydrophobic or water fearing. This allows the membrane to orient itself with the polar heads facing the interstitial fluid found outside the cell and the internal fluid of the cell.

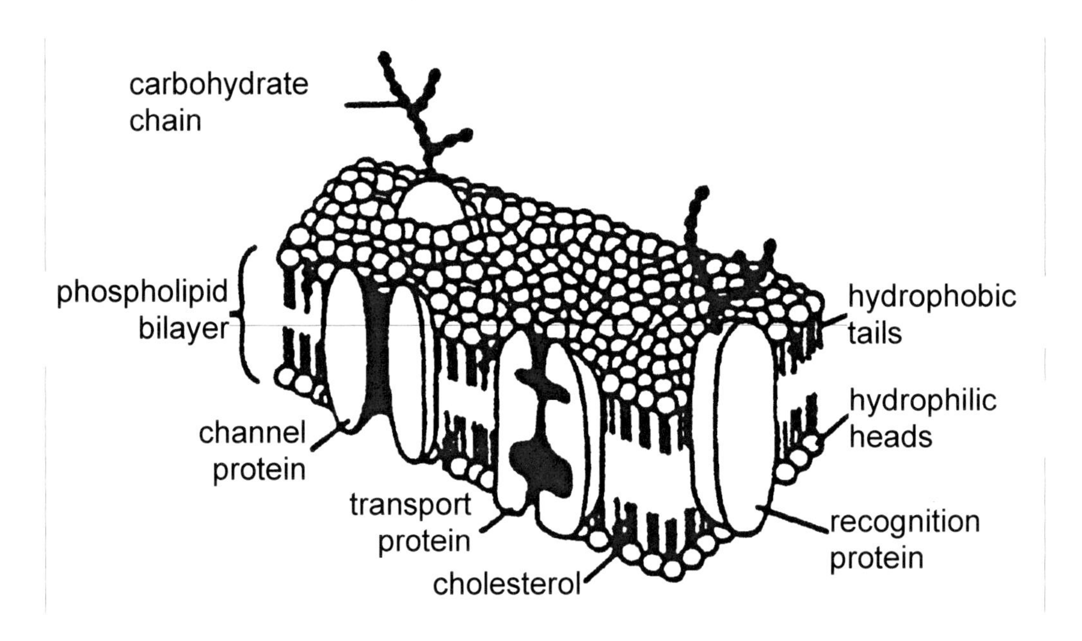

Steroids are insoluble in water and are composed of a carbon skeleton consisting of four inter-connected rings. An important steroid is cholesterol, which is the precursor from which other steroids are synthesized. Hormones, including cortisone, testosterone, estrogen, and progesterone are steroids. Their insolubility keeps them from dissolving in body fluids.

Proteins compose about fifty percent of the dry weight of animals and bacteria. Proteins function in structure and aid in support (connective tissue, hair, feathers, quills), storage of amino acids (albumin in eggs, casein in milk), transport of substances (hemoglobin), hormones to coordinate body activities (insulin), membrane receptor proteins, contraction (muscles, cilia, flagella), body defense (antibodies), and as enzymes to speed up chemical reactions.

All proteins are made of twenty **amino acids**. An amino acid contains an amino group and an acid group. The radical group varies and defines the amino acid. Amino acids form through condensation reactions with the removal of water. The bond that is formed between two amino acids is called a peptide bond. Polymers of amino acids are called polypeptide chains. An analogy can be drawn between the twenty amino acids and the alphabet. Millions of words can be formed using an alphabet of only twenty-six letters. This diversity is also possible using only twenty amino acids. This results in the formation of many different proteins, whose structure defines the function.

There are four levels of protein structure: primary, secondary, tertiary, and quaternary.

Primary structure is the protein's unique sequence of amino acids. A slight change in primary structure can affect a protein's conformation and its ability to function. **Secondary structure** is formed by the coils and folds of polypeptide chains. The coils and folds are the result of hydrogen bonds along the polypeptide backbone. The secondary structure is either in the form of an alpha helix or a pleated sheet. The alpha helix is a coil held together by hydrogen bonds. A pleated sheet is the polypeptide chain folding back and forth. The hydrogen bonds between parallel regions hold it together. **Tertiary structure** is formed by bonding between the side chains of the amino acids. Disulfide bridges are created when two sulfhydryl groups on the amino acids bond together to form a strong covalent bond. **Quaternary structure** is the overall structure of the protein from the aggregation of two or more polypeptide chains. An example of this is hemoglobin. Hemoglobin consists of two kinds of polypeptide chains.

Nucleic acids consist of DNA (deoxyribonucleic acid) and RNA (ribonucleic acid). Nucleic acids contain the instructions for the amino acid sequence of proteins and the instructions for replicating. The monomer of nucleic acids is called a nucleotide. A nucleotide consists of a 5 carbon sugar, (deoxyribose in DNA, ribose in RNA), a phosphate group, and a nitrogenous base. The base sequence codes for the instructions. There are five bases: adenine, thymine, cytosine, guanine, and uracil. Uracil is found only in RNA and replaces the thymine. A summary of nucleic acid structure can be seen in the table below:

	SUGAR	PHOSPHATE	BASES
DNA	deoxy-ribose	present	adenine, thymine, cytosine, guanine
RNA	ribose	present	adenine, uracil, cytosine, guanine

Due to the molecular structure, adenine will always pair with thymine in DNA or uracil in RNA. Cytosine always pairs with guanine in both DNA and RNA. This allows for the symmetry of the DNA molecule seen below.

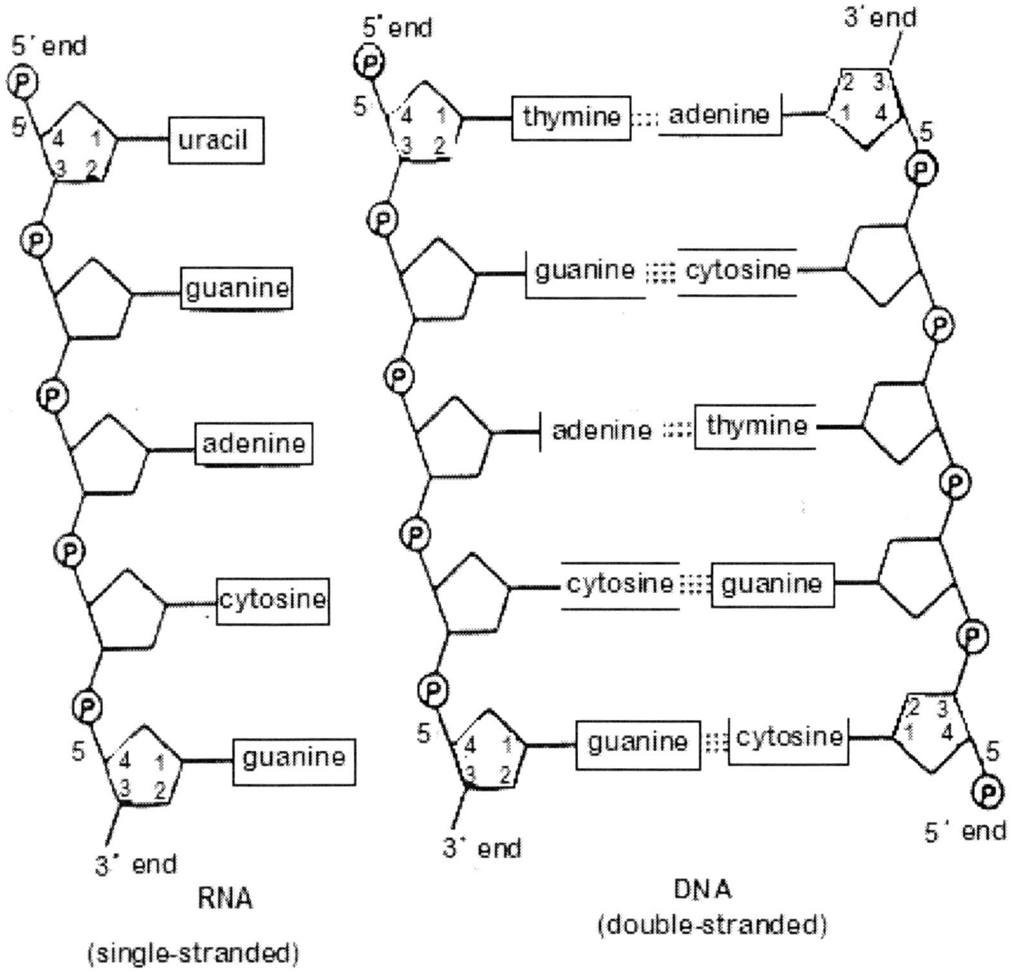

RNA
(single-stranded)

DNA
(double-stranded)

Adenine and thymine (or uracil) are linked by two hydrogen bonds and cytosine and guanine are linked by three hydrogen bonds. Weak hydrogen bonds make it easier for the DNA strands to separate prior to replication and transcription. The DNA molecule is called a double helix due to its twisted ladder shape.

Skill 6.2 Demonstrate knowledge of the structure and function of the parts of a cell and cellular organelles (e.g., cell membrane, nucleus, mitochondria)

ORGANELLES:

The organization of living systems builds by levels from small to increasingly more large and complex. All aspects, whether it be a cell or an ecosystem, have the same requirements to sustain life. Life is organized from simple to complex in the following way:

Organelles make up **cells** which make up **tissues** which make up **organs**. Groups of organs make up **organ systems**. Organ systems work together to provide life for the **organism.**

Parts of Eukaryotic Cells

1. **Nucleus** - The brain of the cell. The nucleus contains:

chromosomes- DNA, RNA and proteins tightly coiled to conserve space while providing a large surface area.
chromatin - loose structure of chromosomes. Chromosomes are called chromatin when the cell is not dividing.
nucleoli - where ribosomes are made. These are seen as dark spots in the nucleus.
nuclear membrane - contains pores which let RNA out of the nucleus. The nuclear membrane is continuous with the endoplasmic reticulum which allows the membrane to expand or shrink if needed.

2. **Ribosomes** - the site of protein synthesis. Ribosomes may be free floating in the cytoplasm or attached to the endoplasmic reticulum. There may be up to a half a million ribosomes in a cell, depending on how much protein is made by the cell.

3. **Endoplasmic Reticulum** - These are folded and provide a large surface area. They are the "roadway" of the cell and allow for transport of materials. The lumen of the endoplasmic reticulum helps to keep materials out of the cytoplasm and headed in the right direction. The endoplasmic reticulum is capable of building new membrane material. There are two types:

Smooth Endoplasmic Reticulum - contain no ribosomes on their surface.

Rough Endoplasmic Reticulum - contain ribosomes on their surface. This form of ER is abundant in cells that make many proteins, like in the pancreas, which produces many digestive enzymes.

4. **Golgi Complex or Golgi Apparatus** - This structure is stacked to increase surface area. The Golgi Complex functions to sort, modify and package molecules that are made in other parts of the cell. These molecules are either sent out of the cell or to other organelles within the cell.

5. **Lysosomes** - found mainly in animal cells. These contain digestive enzymes that break down food, substances not needed, viruses, damaged cell components, and eventually the cell itself. It is believed that lysosomes are responsible for the aging process.

6. **Mitochondria** - large organelles that make ATP to supply energy to the cell. Muscle cells have many mitochondria because they use a great deal of energy. The folds inside the mitochondria are called cristae. They provide a large surface where the reactions of cellular respiration occur. Mitochondria have their own DNA and are capable of reproducing themselves if a greater demand is made for additional energy.

7. **Plastids** - found in photosynthetic organisms only. They are similar to the mitochondria due to their double membrane structure. They also have their own DNA and can reproduce if increased capture of sunlight becomes necessary. There are several types of plastids:

Chloroplasts - green, function in photosynthesis. They are capable of trapping sunlight.
Chromoplasts - make and store yellow and orange pigments; they provide color to leaves, flowers and fruits.
Amyloplasts - store starch and are used as a food reserve. They are abundant in roots like potatoes.

8. **Cell Wall** - found in plant cells only, it is composed of cellulose and fibers. It is thick enough for support and protection, yet porous enough to allow water and dissolved substances to enter. Cell walls are cemented to each other.

9. **Vacuoles** - hold stored food and pigments. Vacuoles are very large in plants. This is allows them to fill with water in order to provide turgor pressure. Lack of turgor pressure causes a plant to wilt.

10. **Cytoskeleton** - composed of protein filaments attached to the plasma membrane and organelles. They provide a framework for the cell and aid in cell movement. They constantly change shape and move about. Three types of fibers make up the cytoskeleton:

Microtubules - largest of the three; makes up cilia and flagella for locomotion. Flagella grow from a basal body. Some examples are sperm cells, and tracheal cilia. Centrioles are also composed of microtubules. They form the spindle fibers that pull the chromosomes apart to form two cells during cell division. Centrioles are not found in the cells of higher plants.

Intermediate Filaments - they are smaller than microtubules but larger than microfilaments. They help the cell to keep its shape.

Microfilaments - smallest of the three, they are made of actin and small amounts of myosin (like in muscle cells). They function in cell movement such as cytoplasmic streaming, endocytosis, and ameboid movement. This structure pinches the two cells apart after cell division, forming two cells.

STRUCTURE AND FUNCTION:

The structure of the cell is often related to the cell's function. Root hair cells differ from flower stamens or leaf epidermal cells. They all have different functions.

Animal cells – The nucleus is a round body inside the cell. It controls the cell's activities. The nuclear membrane contains threadlike structures called chromosomes. The genes are units found on chromosomes that control cell activities found in the nucleus. The cytoplasm has many structures in it. Vacuoles contain the food for the cell. Other vacuoles contain waste materials. Animal cells differ from plant cells because they have centrioles.

Plant cells – have cell walls. A cell wall differs from cell membranes. The cell membrane is very thin and is a part of the cell. The cell wall is thick and is a nonliving part of the cell. Chloroplasts are bundles of chlorophyll.

Single cells – A single celled eukaryotic organism is called a **protist.** When you look under a microscope, the animal-like protists are called **protozoans.** They do not have chloroplasts. They are usually classified by the way they move for food. Amoebas engulf other protists by flowing around and over them. The paramecium has a hair like structure that allows it to move back and forth like tiny oars searching for food. The euglena is an example of a protozoan that moves with a tail-like structure called a flagellum. Plant-like protists have cell walls and float in the water.

Bacteria are the simplest microorganisms. A bacterial cell is surrounded by a cell wall, but there is no nucleus inside the cell. Most bacteria do not contain chlorophyll so they do not make their own food. The classification of bacteria is by shape. Cocci are round, bacilli are rod-shaped, and spirilla are spiral shaped.

Skill 6.3 Demonstrate an understanding of the processes by which cells transport materials across cell membranes (e.g., osmosis, diffusion, active transport)

In order to understand cellular transport, it is important to know about the structure of the cell membrane. All organisms contain cell membranes because they regulate the flow of materials into and out of the cell. The current model for the cell membrane is the Fluid Mosaic Model because of the ability of lipids and proteins to move and change places, giving the membrane fluidity.

Cell membranes have the following characteristics:

1. They are made of phospholipids that have polar, charged heads with a phosphate group that is hydrophilic (water loving) and two nonpolar lipid tails that are hydrophobic (water fearing). This allows the membrane to orient itself with the polar heads facing the fluid inside and outside the cell and the hydrophobic lipid tails sandwiched in between. Each individual phospholipid is called a micelle.

2. They contain proteins embedded inside (integral proteins) and proteins on the surface (peripheral proteins). These proteins may act as channels for transport, may contain enzymes, may act as receptor sites, may act to stick cells together or may attach to the cytoskeleton to give the cell shape.

3. They contain cholesterol, which alters the fluidity of the membrane.

4. They contain oligosaccharides (small carbohydrate polymers) on the outside of the membrane. These act as markers that help distinguish one cell from another.

5. They contain receptors made of glycoproteins that can attach to certain molecules, like hormones.

Cell transport is necessary to maintain homeostasis, or balance of the cell with its external environment. Cell membranes are selectively permeable, which is the key to transport. Not all molecules may pass through easily. Some molecules require energy or carrier molecules and may only cross when needed.

Passive transport does not require energy and moves the material with the concentration gradient (high to low). Small molecules may pass through the membrane in this manner. Two examples of passive transport include diffusion and osmosis. Diffusion is the ability of molecules to move from areas of high concentration to areas of low concentration. It normally involves small uncharged particles like oxygen. Osmosis is simply the diffusion of water across a semi-permeable membrane. Osmosis may cause cells to swell or shrink, depending on the internal and external environments. The following terms are used in relation of the cell to the environment.

Isotonic - water concentration is equal inside and outside the cell. Net movement in either direction is basically equal.

Hypertonic - "hyper" refers to the amount of dissolved particles. The more particles in a solution, the lower its water concentration. Therefore, when a cell is hypertonic to its environment, there is more water outside the cell than inside. Water will move into the cell and the cell will swell. If the environment is hypertonic to the cell, there is more water inside the cell. Water will move out of the cell and the cell will shrink.

Hypotonic - "hypo" again refers to the amount of dissolved particles. The less particles in a solution, the higher its water concentration. When a cell is hypotonic to its environment, there is more water inside the cell than outside. Water will move out of the cell and the cell will shrink. If the environment is hypotonic to the cell, there is more water outside the cell than inside. Water will move into the cell and the cell will swell.

The facilitated diffusion mechanism does not require energy, but does require a carrier protein. An example would be insulin, which is needed to carry glucose into the cell.

Active transport requires energy. The energy for this process comes from either ATP or an electrical charge difference. Active transport may move materials either with or against a concentration gradient. Some examples of active transport are:

1. Calcium pumps - actively pump calcium outside of the cell and are important in nerve and muscle transmission.

2. Stomach acid pump - exports hydrogen ions to lower the pH of the stomach and increase acidity.

3. Sodium-Potassium pump - maintains an electrical difference across the cell. This is useful in restoring ion balance so nerves can continue to function. It exchanges sodium ions for potassium ions across the plasma membrane in animal cells.

Active transport involves a membrane potential which is a charge on the membrane. The charge works like a magnet and may cause transport proteins to alter their shape, thus allowing substances in or out of the cell.

The transport of large molecules depends on the fluidity of the membrane that is controlled by cholesterol in the membrane. Exocytosis is the release of large particles by the vesicles fusing with the plasma membrane. In the process of endocytosis, the cell takes in macromolecules and particulate matter by forming vesicles derived from the plasma membrane. There are three types of endocytosis in animal cells. Phagocytosis is when a particle is engulfed by pseudopodia and packaged in a vacuole. In pinocytosis, the cell takes in extracellular fluid in small vesicles. Receptor-mediated endocytosis is when the membrane vesicles bud inward to allow a cell to take in large amounts of certain substances. The vesicles have proteins with receptors that are specific for the substance.

Skill 6.4 Compare and contrast viruses and bacteria, prokaryotic and eukaryotic cells, and plant and animal cells

The cell is the basic unit of all living things. There are three types of cells. They are prokaryotes, eukaryotes, and archaea. Archaea have some similarities with prokaryotes, but are as distantly related to prokaryotes as prokaryotes are to eukaryotes.

PROKARYOTES

Prokaryotes (Monera) consist only of bacteria and cyanobacteria (formerly known as blue-green algae). Bacteria are then further divided into 19 groups. Prokaryote cells have no defined nucleus or nuclear membrane. The DNA, RNA, and ribosomes float freely within the cell. The cytoplasm has a single chromosome condensed to form a nucleoid. Prokaryotes have a thick cell wall made up of amino sugars (glycoproteins). This is for protection, to give the cell shape, and to keep the cell from bursting. It is the cell wall of bacteria that is targeted by the antibiotic penicillin. Penicillin works by disrupting the cell wall, thus killing the cell.

The cell wall surrounds the cell membrane (plasma membrane). The cell membrane consists of a lipid bilayer that controls the passage of molecules in and out of the cell. Some prokaryotes have a capsule made of polysaccharides that surrounds the cell wall for extra protection from higher organisms.

Many bacterial cells have appendages used for movement called flagella. Some cells also have pili, which are a protein strand used for attachment of the bacteria. Pili may also be used for sexual conjugation (where the DNA from one bacterial cell is transferred to another bacterial cell).

Prokaryotes are the most numerous and widespread organisms on earth. Bacteria were most likely the first cells and date back in the fossil record to 3.5 billion years ago. Their ability to adapt to the environment allows them to thrive in a wide variety of habitats.

EUKARYOTES

Eukaryotic cells are found in protists, fungi, plants, and animals. Most eukaryotic cells are larger than prokaryotic cells. They contain many organelles, which are membrane bound areas for specific functions. Their cytoplasm contains a cytoskeleton that provides a protein framework for the cell. The cytoplasm also supports the organelles and contains the ions and molecules necessary for cell function. The cytoplasm is contained by the plasma membrane. The plasma membrane allows molecules to pass in and out of the cell. The membrane can bud inward to engulf outside material in a process called endocytosis. Exocytosis is a secretory mechanism, the reverse of endocytosis.

The most significant differentiation between prokaryotes and eukaryotes is that eukaryotes have a nucleus. Although bacteria and fungi may cause disease, they are also beneficial for use as medicines and food. Penicillin is derived from a fungus that is capable of destroying the cell wall of bacteria. Most antibiotics work in this way. Viral diseases have been fought through the use of vaccination, where a small amount of the virus is introduced so the immune system is able to recognize it upon later infection. Antibodies are more quickly manufactured when the host has had prior exposure. Viruses are difficult to treat because antibiotics are ineffective against them. That is why doctors do not usually prescribe antibiotics for those who have a cold or the flu—common viral infections. While some yeasts can cause illness, Brewer's yeast is a fungi that humans use to make bread and to ferment wine.

ARCHAEA

There are three kinds of organisms with archaea cells: methanogens are obligate anaerobes that produce methane, halobacteria can live only in concentrated brine solutions, and thermoacidophiles can only live in acidic hot springs.

VIRUSES

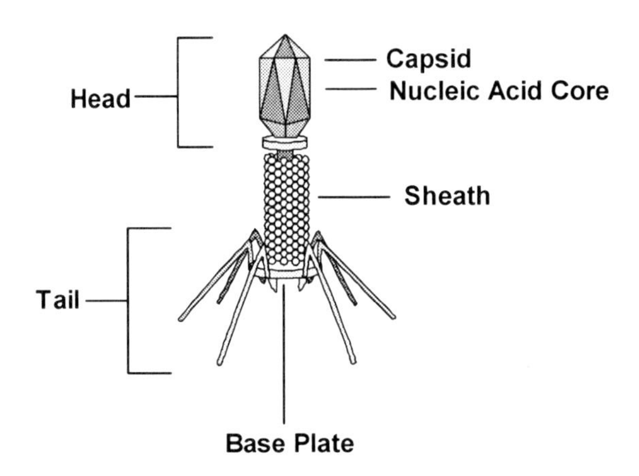

Bacteriophage

All viruses have a head or protein capsid that contains genetic material. This material is encoded in the nucleic acid and can be DNA, RNA, or even a limited number of enzymes. Some viruses also have a protein tail region. The tail aids in binding to the surface of the host cell and penetrating the surface of the host in order to introduce the virus's genetic material.

Other examples of viruses and their structures:

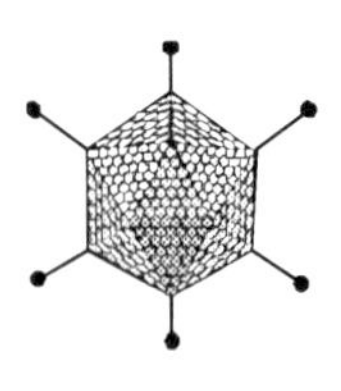

Adenovirus (DNA virus)

Eastern equine encephalitis (RNA virus)

Herpes virus (DNA virus)

HIV retrovirus (RNA virus)

Influenza virus (RNA virus)

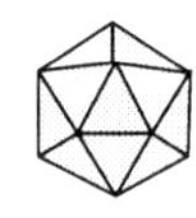

Rotavirus (RNA virus)

Skill 6.5 Demonstrate an understanding of the cell cycle and the processes of mitosis and meiosis

The purpose of cell division is to provide growth and repair in body (somatic) cells and to replenish or create sex cells for reproduction. There are two forms of cell division. **Mitosis** is the division of somatic cells and **meiosis** is the division of sex cells (eggs and sperm). The table below summarizes the major differences between the two processes.

MITOSIS	MEIOSIS
1. Division of somatic cell	1. Division of sex cells
2. Two cells result from each division	2. Four cells or polar bodies result from each division
3. Chromosome number is identical to parent cells.	3. Chromosome number is half the number of parent cells
4. For cell growth and repair	4. Recombinations provide genetic diversity

Some terms to know

gamete - sex cell or germ cell; eggs and sperm.
chromatin - loose chromosomes; this state is found when the cell is not dividing.
chromosome - tightly coiled, visible chromatin; this state is found when the cell is dividing.
homologues - chromosomes that contain the same information. They are of the same length and contain the same genes.
diploid - 2n number; diploid chromosomes are a pair of chromosomes (somatic cells).
haploid - 1n number; haploid chromosomes are a half of a pair (sex cells).

MITOSIS

The cell cycle is the life cycle of the cell. It is divided into two stages; **Interphase** and the **mitotic division** where the cell is actively dividing. Interphase is divided into three steps; G1 (growth) period, where the cell is growing and metabolizing, S period (synthesis) where new DNA and enzymes are being made and the G2 phase (growth) where new proteins and organelles are being made to prepare for cell division. The mitotic stage consists of the stages of mitosis and the division of the cytoplasm.

The stages of mitosis and their events are as follows. Be sure to know the correct order of steps. (IPMAT)

1. Interphase - chromatin is loose, chromosomes are replicated, cell metabolism is occurring. Interphase is technically not a stage of mitosis.

2. Prophase - once the cell enters prophase, it proceeds through the following steps continuously, with no stopping. The chromatin condenses to become visible chromosomes. The nucleolus disappears and the nuclear membrane breaks apart. Mitotic spindles form which will eventually pull the chromosomes apart. They are composed of microtubules. The cytoskeleton breaks down and the spindles are pushed to the poles or opposite ends of the cell by the action of centrioles.

3. Metaphase - kinetechore fibers attach to the chromosomes which causes the chromosomes to line up in the center of the cell (think **m**iddle for **m**etaphase)

4. Anaphase - centromeres split in half and homologous chromosomes separate. The chromosomes are pulled to the poles of the cell, with identical sets at either end.

5. Telophase - two nuclei form with a full set of DNA that is identical to the parent cell. The nucleoli become visible and the nuclear membrane reassembles. A cell plate is visible in plant cells, whereas a cleavage furrow is formed in animal cells. The cell is pinched into two cells. Cytokinesis, or division, of the cytoplasm and organelles occurs.

Meiosis contains the same five stages as mitosis, but is repeated in order to reduce the chromosome number by one half. This way, when the sperm and egg join during fertilization, the haploid number is reached. The steps of meiosis are as follows:

Major function of Meiosis I – chromosomes are replicated; cell becomes haploid
Prophase I - replicated chromosomes condense and pair with homologues. This forms a tetrad. Crossing over (the exchange of genetic material between homologues to further increase diversity) occurs during Prophase I.
Metaphase I - homologous sets attach to spindle fibers after lining up in the middle of the cell.
Anaphase I - sister chromatids remain joined and move to the poles of the cell.
Telophase I - two new cells are formed, chromosome number are now haploid

Major function of Meiosis II - to reduce the chromosome number in half.

Prophase II - chromosomes condense.
Metaphase II - spindle fibers form again, sister chromatids line up in center of cell, centromeres divide and sister chromatids separate.
Anaphase II - separated chromosomes move to opposite ends of cell.
Telophase II - four haploid cells form for each original sperm germ cell. One viable egg cell gets all the genetic information and three polar bodies form with no DNA. The nuclear membrane reforms and cytokinesis occurs.

Skill 6.6 Demonstrate knowledge of abnormal cell structures and processes and their relation to diseases (e.g., cancer)

MUTATIONS THAT LEAD TO ABNORMALTIES

Since it's not a perfect world, mistakes happen. Inheritable changes in DNA are called **mutations**. Mutations may be errors in replication or a spontaneous rearrangement of one or more segments by factors like radioactivity, drugs, or chemicals. The amount of the change is not as critical as where the change is. Mutations may occur on somatic or sex cells. Usually the ones on sex cells are more dangerous since they contain the basis of all information for the developing offspring. Mutations are not always bad. They are the basis of evolution, and if they make a more favorable variation that enhances the organism's survival, then they are beneficial. But, mutations may also lead to abnormalities, birth defects, and even death. There are several types of mutations; let's suppose a normal sequence was as follows:

Normal - A B C D E F

Duplication - one base is repeated. A B C C D E F

Inversion - a segment of the sequence is flipped around. A E D C B F

Deletion - a base is left out. A B C E F

Insertion or Translocation - a segment from another place on the DNA is inserted in the wrong place. A B C R S D E F

Breakage - a piece is lost. A B C (DEF is lost)

Nondisjunction – This occurs during meiosis when chromosomes fail to separate properly. One sex cell may get both chromosomes and another may get none. Depending on the chromosomes involved this may or may not be serious. Offspring end up with either an extra chromosome or are missing one. An example of nondisjunction is Down Syndrome, where three of chromosome #21 are present.

Relationship of unrestricted cell cycle and cancer

The restriction point for cellular division occurs late in the G1 phase of the cell cycle. This is when the decision for the cell to divide is made. If all the internal and external cell systems are working properly, the cell proceeds to replicate. Cells may also decide not to proceed past the restriction point. This non-dividing cell state is called the G0 phase. Many specialized cells remain in this state.

The density of cells also regulates cell division. Density-dependent inhibition is when the cells crowd one another and consume all the nutrients, therefore halting cell division. Cancer cells do not respond to density-dependent inhibition. They divide excessively and invade other tissues. As long as there are nutrients, cancer cells are "immortal."

COMPETENCE 7.0 UNDERSTAND THE STRUCTURE, ORGANIZATION, AND BASIC LIFE FUNCTIONS OF ORGANISMS

Skill 7.1 Demonstrate knowledge of the levels of organization in organisms (i.e., cells, tissues, organs, and systems) and the relationship of organs and organ systems to each other and to the organism as a whole

Life is highly organized. The organization of living systems builds on levels from small to increasingly more large and complex. All aspects, whether it is a cell or an ecosystem, have the same requirements to sustain life. Life is organized from simple to complex in the following way:

Atoms→molecules→organelles→cells→tissues→organs→organ systems→organism

Groups of related organs are organ systems. Organ systems consist of organs working together to perform a common function. The commonly recognized organ systems of animals include the reproductive system, nervous system, circulatory system, respiratory system, lymphatic system (immune system), endocrine system, excretory system, muscular system, digestive system, integumentary system, and skeletal system. In addition, organ systems are interconnected and a single system rarely works alone to complete a task.

One obvious example of the interconnectedness of organ systems is the relationship between the circulatory and respiratory systems. As blood circulates through the organs of the circulatory systems, it is re-oxygenated in the lungs of the respiratory system. Another example is the influence of the endocrine system on other organ systems. Hormones released by the endocrine system greatly influence processes of many organ systems including the nervous and reproductive systems.

In addition, bodily response to infection is a coordinated effort of the lymphatic (immune system) and circulatory systems. The lymphatic system produces specialized immune cells, filters out disease-causing organisms, and removes fluid waste from in and around tissue. The lymphatic system utilizes capillary structures of the circulatory system and interacts with blood cells in a coordinated response to infection.

Finally, the muscular and skeletal systems are closely related. Skeletal muscles attach to the bones of the skeleton and drive movement of the body.

Skill 7.2 Analyze the relationship between structure and function in different systems for various types of organisms (e.g., respiratory, digestive)

The function of different systems in organisms from bacteria to humans dictates system structure. The basic principle that "form follows function" applies to all organismal systems. We will discuss a few examples to illustrate this principle. Keep in mind that we can relate the structure and function of all organismal systems.

Mitochondria, subcellular organelles present in eukaryotic cells, provide energy for cell functions. Much of the energy-generating activity takes place in the mitochondrial membrane. To maximize this activity, the mitochondrial membrane has many folds to pack a relatively large amount of membrane into a small space.

Bacterial cells maintain a high surface area to volume ratio to maximize contact with the environment and allow for exchange of nutrients and waste products. Bacterial cells achieve this high ratio by maintaining a small internal volume by cell division.

Skeletal System - The skeletal system functions in support. Vertebrates have an endoskeleton, with muscles attached to bones. Skeletal proportions are controlled by area to volume relationships. Body size and shape is limited due to the forces of gravity. Surface area is increased to improve efficiency in all organ systems. Finally, the structure of the skeletal systems of different animals varies based on the animal's method of movement. For example, the honeycombed structure of bird bones provides a lightweight skeleton of great strength to accommodate flight. The bones of the human skeletal system are dense, strong, and aligned in such a way as to allow walking on two legs in an upright position.

Muscular System – Its function is for movement. There are three types of muscle tissue. Skeletal muscle is voluntary. These muscles are attached to bones. Smooth muscle is involuntary. It is found in organs and enable functions such as digestion and respiration. Cardiac muscle is a specialized type of smooth muscle.

Nervous System - The neuron is the basic unit of the nervous system. It consists of an axon, which carries impulses away from the cell body, the dendrite, which carries impulses toward the cell body and the cell body, which contains the nucleus. Synapses are spaces between neurons. Chemicals called neurotransmitters are found close to the synapse. The myelin sheath, composed of Schwann cells, covers the neurons and provides insulation.

Digestive System - The function of the digestive system is to break down food and absorb it into the blood stream where it can be delivered to all cells of the body for use in cellular respiration. As animals evolved, digestive systems changed from simple absorption to a system with a separate mouth and anus, capable of allowing the animal to become independent of a host.

Respiratory System - This system functions in the gas exchange of oxygen (needed) and carbon dioxide (waste). It delivers oxygen to the bloodstream and picks up carbon dioxide for release out of the body. Simple animals diffuse gases from and to their environment. Gills allow aquatic animals to exchange gases in a fluid medium by removing dissolved oxygen from the water. Lungs maintain a fluid environment for gas exchange in terrestrial animals.

Circulatory System - The function of the circulatory system is to carry oxygenated blood and nutrients to all cells of the body and return carbon dioxide waste to be expelled from the lungs. Animals evolved from an open system to a closed system with vessels leading to and from the heart.

The **cardiovascular system** of animals has many specialized structures that help achieve the function of delivering blood to all parts of the body. The heart has four chambers for the delivery and reception of blood. The blood vessels vary in size to accommodate the necessary volume of blood. For example, vessels near the heart are large to accommodate large amounts of blood and vessels in the extremities are very small to limit the amount of blood delivered.

Skill 7.3 Compare and contrast how various organisms carry out basic life processes (e.g., maintaining homeostasis, reproduction, growth)

Members of the six different kingdoms of the classification system of living organisms often differ in their basic life functions. Here we compare and analyze how members of the six kingdoms obtain nutrients, excrete waste, and reproduce.

Bacteria are prokaryotic, single-celled organisms that lack cell nuclei. The different types of bacteria obtain nutrients in a variety of ways. Most bacteria absorb nutrients from the environment through small channels in their cell walls and membranes (chemotrophs) while some perform photosynthesis (phototrophs). Chemoorganotrophs use organic compounds as energy sources while chemolithotrophs can use inorganic chemicals as energy sources. Depending on the type of metabolism and energy source, bacteria release a variety of waste products (e.g. alcohols, acids, carbon dioxide) to the environment through diffusion. All bacteria reproduce through binary fission (asexual reproduction) producing two identical cells. Bacteria reproduce very rapidly, dividing or doubling every twenty minutes in optimal conditions. Asexual reproduction does not allow for genetic variation, but bacteria achieve genetic variety by absorbing DNA from ruptured cells and conjugating or swapping chromosomal or plasmid DNA with other cells.

Animals are multicellular, eukaryotic organisms. All animals obtain nutrients by eating food (ingestion). Different types of animals derive nutrients from eating plants, other animals, or both. Animal cells perform respiration that converts food molecules, mainly carbohydrates and fats, into energy. The excretory systems of animals, like animals themselves, vary in complexity. Simple invertebrates eliminate waste through a single tube, while complex vertebrates have a specialized system of organs that process and excrete waste.

Most animals, unlike bacteria, exist in two distinct sexes. Members of the female sex give birth or lay eggs. Some less developed animals can reproduce asexually. For example, flatworms can divide in two and some unfertilized insect eggs can develop into viable organisms. Most animals reproduce sexually through various mechanisms. For example, aquatic animals reproduce by external fertilization of eggs, while mammals reproduce by internal fertilization. More developed animals possess specialized reproductive systems and cycles, that facilitate reproduction and promote genetic variation.

Plants, like animals, are multi-cellular, eukaryotic organisms. Plants obtain nutrients from the soil through their root systems and convert sunlight into energy through photosynthesis. Many plants store waste products in vacuoles or organs (e.g. leaves, bark) that are discarded. Some plants also excrete waste through their roots.

More than half of the plant species reproduce by producing seeds from which new plants grow. Depending on the type of plant, flowers or cones produce seeds. Other plants reproduce by spores, tubers, bulbs, buds, and grafts. The flowers of flowering plants contain the reproductive organs. Pollination is the joining of male and female gametes that is often facilitated by movement by wind or animals.

Fungi are eukaryotic, mostly multi-cellular organisms. All fungi are heterotrophs, obtaining nutrients from other organisms. More specifically, most fungi obtain nutrients by digesting and absorbing nutrients from dead organisms. Fungi secrete enzymes outside of their body to digest organic material and then absorb the nutrients through their cell walls.

Most fungi can reproduce asexually and sexually. Different types of fungi reproduce asexually by mitosis, budding, sporification, or fragmentation. Sexual reproduction of fungi is different from sexual reproduction of animals. The two mating types of fungi are plus and minus, not male and female. The fusion of hyphae, the specialized reproductive structure in fungi, between plus and minus types produces and scatters diverse spores.

Protists are eukaryotic, single-celled organisms. Most protists are heterotrophic, obtaining nutrients by ingesting small molecules and cells and digesting them in vacuoles. All protists reproduce asexually by either binary or multiple fission. Like bacteria, protists achieve genetic variation by exchange of DNA through conjugation.

Skill 7.4 Compare and contrast sources of energy and matter for different organisms and how various organisms obtain, store, and use energy and matter

Sources of energy

The ultimate source of energy for most ecosystems is solar radiation. Primary producers are usually the organisms in an ecosystem that can convert light energy into chemical energy. Most primary producers are photosynthetic. Photosynthetic primary producers include algae, plants, and many species of bacteria. All other organisms in an ecosystem depend on primary producers to provide energy.

The main primary producers in terrestrial ecosystems are plants. In limnetic (deep-water) zone lake and pond ecosystems and open ocean ecosystems, algae and photosynthetic bacteria are the most important primary producers. In littoral (shallow water, near-shore) zone freshwater and ocean ecosystems, the main primary producers are aquatic plants and multicellular algae. Finally, one notable exception to the photosynthetic organism as primary producer rule are ecosystems near hot water vents on the deep-sea floor. Because solar energy is unavailable, chemoautotrophic bacteria that can oxidize hydrogen sulfide are the primary producers.

Obtaining energy

All organisms can be classes by the manner in which they obtain energy: chemoautotrophs, photoautotrophs, and heterotrophs.

Chemoautotrophs- These organisms are able to obtain energy via the oxidation of inorganic molecules (i.e., hydrogen gas and hydrogen sufide) or methane. This process is known as chemosynthesis. Most chemoautotrophs are bacteria or archaea that thrive in oxygen-poor environments, such as deep sea vents.

Photoautotrophs- Instead of obtaining energy from simple inorganic compounds like the chemoautotrophs, organisms of this type receive energy from sunlight. They employ the process of photosynthesis to create sugar from light, carbon dioxide and water. Most higher plants and algae as well as some bacteria and protists are photoautotrophs.

Heterotrophs- Any organism that requires organic molecules for as its source of energy is a heterotroph. These organisms are consumers in the food chain and must obtain nutrition from autotrophs or other heterotrophs. All animals are heterotrophs, as are some fungi and bacteria.

Structures used to store food

Vacuoles are one of the most common storage compartments used by simple organisms and plants. Though we commonly think of vacuoles as simply the organelle employed by plant cells to maintain turgor pressure, they actually have a variety of functions. For example, in budding yeast cells, vacuoles are used as storage compartments for amino acids. When the yeast cells are deprived of food, proteins are consumed within the vacuoles. This process is known as autophagy. Additionally, protists and macrophages use vacuoles to hold food when they engage in phagocytosis (the cellular intake of large molecules or other cells).

Plants have evolved various methods to store excess food. While some store extra glucose, many plants store starch in their roots and stems. Seeds are often "packaged" with additional stored food in the form of both sugar and starch. Many common examples of such plant "storage devices" are exploited as food by humans. For instance, carrots are large roots packed with the plant's extra food. Most fruits, nuts, and edible seeds also contain many calories intended to nurture the next generation of plants.

In many animals, adipose (fat) tissue is used to store extra metabolic energy for long periods. Excess calories are metabolized by the liver into fat. This, along with dietary fat, is stored by adipocytes. When this energy is needed by the body, the stored fat can supply be broken down to supply fatty acids and glycerol. Glycerol can be converted to glucose and used as a source of energy for many cells in the body, while fatty acids are especially needed by the heart and skeletal muscle. The storage and use of this fat is under the control of several hormones including insulin, glucagons, and epinephrine.

Skill 7.5 Categorize organisms based on methods of reproduction and offspring

The obvious advantage of asexual reproduction is that it does not require a partner. This is a huge advantage for organisms, such as the hydra, which do not move around. Not having to move around to reproduce also allows organisms to conserve energy. Asexual reproduction also tends to be faster. There are disadvantages, as in the case of regeneration, in plants if the plant is not in good condition or in the case of spore-producing plants, if the surrounding conditions are not suitable for the spores to grow. As asexual reproduction produces only exact copies of the parent organism, it does not allow for genetic variation, which means that mutations, or weaker qualities, will always be passed on. This can also be detrimental to a species well-adapted to a particular environment when the conditions of that environment change suddenly. On the whole, asexual reproduction is more reliable because it requires fewer steps and less can go wrong.

Sexual reproduction shares genetic information between gametes, thereby producing variety in the species. This can result in a better species with an improved chance of survival. There is the disadvantage that sexual reproduction requires a partner, which in turn with many organisms requires courtship, finding a mate, and mating. Another disadvantage is that sexually reproductive organisms require special mechanisms.

Skill 7.6 Analyze behavioral responses to external stimuli in a variety of organisms

Response to stimuli is one of the key characteristics of any living thing. Any detectable change in the internal or external environment (the stimulus) may trigger a response in an organism. Just like physical characteristics, organisms' responses to stimuli are adaptations that allow them to better survive. While these responses may be more noticeable in animals that can move quickly, all organisms are actually capable of responding to changes.

Single celled organisms
These organisms are able to respond to basic stimuli such as the presence of light, heat, or food. Changes in the environment are typically sensed via cell surface receptors. These organisms may respond to such stimuli by making changes in internal biochemical pathways or initiating reproduction or phagocytosis. Those capable of simple motility, using flagella for instance, may respond by moving toward food or away from heat.

Plants
Plants typically do not possess sensory organs and so individual cells recognize stimuli through a variety of pathways. When many cells respond to stimuli together, a response becomes apparent. Logically then, the responses of plants occur on a rather longer timescale than those of animals. Plants are capable of responding to a few basic stimuli including light, water and gravity. Some common examples include the way plants turn and grow toward the sun, the sprouting of seeds when exposed to warmth and moisture, and the growth of roots in the direction of gravity.

Animals
Lower members of the animal kingdom have responses similar to those seen in single celled organisms. However, higher animals have developed complex systems to detect and respond to stimuli. The nervous system, sensory organs (eyes, ears, skin, etc), and muscle tissue all allow animals to sense and quickly respond to changes in their environment. As in other organisms, many responses to stimuli in animals are involuntary. For example, pupils dilate in response to the reduction of light. Such reactions are typically called reflexes. However, many animals are also capable of voluntary response. In many animal species, voluntary reactions are instinctual. For instance, a zebra's response to a lion is a *voluntary* one, but, *instinctually*, it will flee quickly as soon as the lion's presence is sensed. Complex responses, which may or may not be instinctual, are typically termed behavior. An example is the annual migration of birds when seasons change. Even more complex social behavior is seen in animals that live in large groups.

COMPETENCE 8.0 UNDERSTAND THE BASIC PRINCIPLES OF GENETICS AND EVOLUTION

Skill 8.1 Relate the structure and behavior of chromosomes during meiosis to hereditary patterns

MEIOSIS/BEHAVIOR OF CHROMOSOMES:

Meiosis contains the same five stages as mitosis, but is repeated in order to reduce the chromosome number by one half. This way, when the sperm and egg join during fertilization, the haploid number is reached. The steps of meiosis are as follows:

Major function of Meiosis I - chromosomes are replicated; cells become haploid

Prophase I - replicated chromosomes condense and pair with homologues. This forms a tetrad. Crossing over (the exchange of genetic material between homologues to further increase diversity) occurs during Prophase I.
Metaphase I - homologous sets attach to spindle fibers after lining up in the middle of the cell.
Anaphase I - sister chromatids remain joined and move to the poles of the cell.
Telophase I - two new cells are formed, chromosome number is now haploid.

Major function of Meiosis II - to reduce the chromosome number in half.

Prophase II - chromosomes condense.
Metaphase II - spindle fibers form again, sister chromatids line up in center of cell, centromeres divide and sister chromatids separate.
Anaphase II - separated chromosomes move to opposite ends of cell.
Telophase II - four haploid cells form for each original sperm germ cell. One viable egg cell gets all the genetic information and three polar bodies form with no DNA. The nuclear membrane reforms and cytokinesis occurs.

STRUCTURE OF CHROMOSOMES:

Chromosomes are the physical structures found in every cell that carry the genetic information of an organism and function in the transmission of hereditary information. Each chromosome contains a sequence of genes each with a specific locus. A locus is the position a given gene occupies on a chromosome. Each gene consists of a sequence of DNA that dictates a particular characteristic of an organism. Separating the genes on a chromosome are regions of DNA that do not code for proteins or other cellular products, but may function in the regulation of coding regions.

HEREDITARY PATTERNS:

Mendel formed three laws:

Law of dominance - in a pair of alleles, one trait may cover up the allele of the other trait. Example: brown eyes are dominant to blue eyes.

Law of segregation - only one of the two possible alleles from each parent is passed on to the offspring from each parent. (During meiosis, the haploid number insures that half the sex cells get one allele, half get the other).

Law of independent assortment - alleles sort independently of each other. (Many combinations are possible depending on which sperm ends up with which egg. Compare this to the many combinations of hands possible when dealing a deck of cards).

Nondisjunction – This occurs during meiosis when chromosomes fail to separate properly. One sex cell may get both chromosomes and another may get none. Depending on the chromosomes involved this may or may not be serious. Offspring end up with either an extra chromosome or are missing one. An example of nondisjunction is Down Syndrome, where three of chromosome #21 are present.

Skill 8.2 Demonstrate knowledge of the principles of inheritance (e.g., dominant, recessive, and sex-linked traits, DNA) and apply probability methods to determine genotype and phenotype frequencies

PRINCIPLES OF INHERITANCE:

SOME DEFINITIONS TO KNOW

Dominant - the stronger of the two traits. If a dominant gene is present, it will be expressed. Dominant traits are represented by a capital letter.

Recessive - the weaker of the two traits. In order for the recessive gene to be expressed, there must be two recessive genes present. Recessive traits are represented by a lower case letter.

Homozygous - (purebred) having two of the same genes present; an organism may be homozygous dominant with two dominant genes or homozygous recessive with two recessive genes.

Heterozygous - (hybrid) having one dominant gene and one recessive gene. The dominant gene will be expressed due to the Law of Dominance.

Genotype - the genes the organism has. Genes are represented with letters. AA, Bb, and tt are examples of genotypes.

Phenotype - how the trait is expressed in an organism. Blue eyes, brown hair, and red flowers are examples of phenotypes.

Incomplete dominance - neither gene masks the other; a new phenotype is formed. For example, red flowers and white flowers may have equal strength. A heterozygote (Rr) would have pink flowers. If a problem occurs with a third phenotype, incomplete dominance is occurring.

Codominance - genes may form new phenotypes. The ABO blood grouping is an example of co-dominance. A and B are of equal strength and O is recessive. Therefore, type A blood may have the genotypes of AA or AO, type B blood may have the genotypes of BB or BO, type AB blood has the genotype A and B, and type O blood has two recessive O genes.

Linkage - genes that are found on the same chromosome usually appear together unless crossing over has occurred in meiosis. (Example - blue eyes and blonde hair)

Lethal alleles - these are usually recessive due to the early death of the offspring. If a 2:1 ratio of alleles is found in offspring, a lethal gene combination is usually the reason. Some examples of lethal alleles include sickle cell anemia, tay-sachs and cystic fibrosis. Usually the coding for an important protein is affected.

Inborn errors of metabolism - these occur when the protein affected is an enzyme. Examples include PKU (phenylketonuria) and albanism.

Polygenic characters - many alleles code for a phenotype. There may be as many as twenty genes that code for skin color. This is why there is such a variety of skin tones. Another example is height. A couple of medium height may have very tall offspring.

Sex linked traits - the Y chromosome found only in males (XY) carries very little genetic information, whereas the X chromosome found in females (XX) carries very important information. Since men have no second X chromosome to cover up a recessive gene, the recessive trait is expressed more often in men. Women need the recessive gene on both X chromosomes to show the trait. Examples of sex linked traits include hemophilia and color-blindness.

Sex influenced traits - traits are influenced by the sex hormones. Male pattern baldness is an example of a sex influenced trait. Testosterone influences the expression of the gene. Mostly men lose their hair due to this trait.

The modern definition of a gene is a unit of genetic information. DNA makes up genes which in turn make up the chromosomes. DNA is wound tightly around proteins in order to conserve space. The DNA/protein combination makes up the chromosome. DNA controls the synthesis of proteins, thereby controlling the total cell activity. DNA is capable of making copies of itself.

DNA

Review of DNA structure:

1. Made of nucleotides; a five carbon sugar, phosphate group and nitrogen base (either adenine, guanine, cytosine or thymine).

2. Consists of a sugar/phosphate backbone which is covalently bonded. The bases are joined down the center of the molecule and are attached by hydrogen bonds which are easily broken during replication.

3. The amount of adenine equals the amount of thymine and the amount of cytosine equals the amount of guanine.

4. The shape is that of a twisted ladder called a double helix. The sugar/phosphates make up the sides of the ladder and the base pairs make up the rungs of the ladder.

DNA Replication

Enzymes control each step of the replication of DNA. The molecule untwists. The hydrogen bonds between the bases break and serve as a pattern for replication. Free nucleotides found inside the nucleus join on to form a new strand. Two new pieces of DNA are formed which are identical. This is a very accurate process. There is only one mistake for every billion nucleotides added. This is because there are enzymes (polymerases) present that proofread the molecule. In eukaryotes, replication occurs in many places along the DNA at once. The molecule may open up at many places like a broken zipper. In prokaryotic circular plasmids, replication begins at a point on the plasmid and goes in both directions until it meets itself.

Base pairing rules are important in determining a new strand of DNA sequence. For example say our original strand of DNA had the sequence as follows:

1. A T C G G C A A T A G C This may be called our sense strand as it contains a sequence that makes sense or codes for something. The complementary strand (or other side of the ladder) would follow base pairing rules (A bonds with T and C bonds with G) and would read:

2. T A G C C G T T A T C G When the molecule opens up and nucleotides join on, the base pairing rules create two new identical strands of DNA

1. A T C G G C A A T A G C	and	2. A T C G G C A A T A G C
T A G C C G T T A T C G		T A G C C G T T A T C G

PROBABILITY METHODS:

Gregor Mendel is recognized as the father of genetics. His work in the late 1800's is the basis of our knowledge of genetics. Although unaware of the presence of DNA or genes, Mendel realized there were factors (now known as genes) that were transferred from parents to their offspring. Mendel worked with pea plants and fertilized the plants himself, keeping track of subsequent generations which led to the Mendelian laws of genetics. Mendel found that two "factors" governed each trait, one from each parent. Traits or characteristics came in several forms, known as alleles. For example, the trait of flower color had white alleles and purple alleles. Mendel formed three laws:

Law of dominance - in a pair of alleles, one trait may cover up the allele of the other trait. Example: brown eyes are dominant to blue eyes.

Law of segregation - only one of the two possible alleles from each parent is passed on to the offspring from each parent. (During meiosis, the haploid number insures that half the sex cells get one allele, half get the other).

Law of independent assortment - alleles sort independently of each other. (Many combinations are possible depending on which sperm ends up with which egg. Compare this to the many combinations of hands possible when dealing a deck of cards).

monohybrid cross - a cross using only one trait.

dihybrid cross - a cross using two traits. More combinations are possible.

Punnet squares - these are used to show the possible ways that genes combine and indicate probability of the occurrence of a certain genotype or phenotype. One parent's genes are put at the top of the box and the other parent at the side of the box. Genes combine on the square just like numbers that are added in addition tables that we learned in elementary school.

Example: Monohybrid Cross - four possible gene combinations

Example: Dihybrid Cross - sixteen possible gene combinations

Skill 8.3 Identify sources of variation in populations (e.g., mutations, sexual reproduction)

Heritable variation is responsible for the individuality of organisms. An individual's phenotype is based on inherited genotype and the surrounding environment. For example, people can alter their phenotypes by lifting weight or diet and exercise. Variation is generated by mutation and sexual recombination. **Mutations** may be errors in replication or a spontaneous rearrangement of one or more segments of DNA.

Mutations contribute a minimal amount of variation in a population. It is the unique **recombination** of existing alleles that causes the majority of genetic differences. Recombination is caused by the crossing over of the parents' genes during meiosis. This results in a unique offspring. With all the possible mating combinations in the world, it is obvious how sexual reproduction is the primary cause of genetic variation.

Skill 8.4 Demonstrate an understanding of the roles of variation, natural selection, and reproductive isolation in adaptation and speciation

Charles Darwin proposed a mechanism for his theory of evolution, which he termed natural selection. Natural selection describes the process by which favorable traits accumulate in a population, changing the population's genetic make-up over time. Darwin theorized that all individual organisms, even those of the same species, are different and those individuals that happen to possess traits favorable for survival would produce more offspring. Thus, in the next generation, the number of individuals with the favorable trait increases and the process continues. Darwin, in contrast to other evolutionary scientists, did not believe that traits acquired during an organism's lifetime (e.g. increased musculature) or the desires and needs of the organism affected evolution of populations. For example, Darwin argued that the evolution of long trunks in elephants resulted from environmental conditions that favored those elephants that possessed longer trunks. The individual elephants did not stretch their trunks to reach food or water and pass on the new, longer trunks to their offspring.

Jean Baptiste Lamarck proposed an alternative mechanism of evolution. Lamarck believed individual organisms developed traits in response to changing environmental conditions and passed on these new, favorable traits to their offspring. For example, Lamarck argued that the trunks of individual elephants lengthen as a result of stretching for scarce food and water, and elephants pass on the longer trunks to their offspring. Thus, in contrast to Darwin's relatively random natural selection, Lamarck believed the mechanism of evolution followed a predetermined plan and depended on the desires and needs of individual organisms.

Different molecular and environmental processes and conditions drive the evolution of populations. The various mechanisms of evolution either introduce new genetic variation or alter the frequency of existing variation.

Mutations, random changes in nucleotide sequence, are a basic mechanism of evolution. Mutations in DNA result from copying errors during cell division, exposure to radiation and chemicals, and interaction with viruses. Simple point mutations, deletions, or insertions can alter the function or expression of existing genes but do not contribute greatly to evolution. On the other hand, gene duplication, the duplication of an entire gene, often leads to the creation of new genes that may contribute to the evolution of a species. Because gene duplication results in two copies of the same gene, the extra copy is free to mutate and develop without the selective pressure experienced by mutated single-copy genes. Gene duplication and subsequent mutation often leads to the creation of new genes. When new genes resulting from mutations lend the mutated organism a reproductive advantage relative to environmental conditions, natural selection and evolution can occur.

Crossing over is the exchange of DNA between a pair of chromosomes during meiosis. Recombination does not introduce new genes into a population, but does affect the expression of genes and the combination of traits expressed by individuals. Thus, crossing over increases the genetic diversity of populations and contributes to evolution by creating new combinations of genes that nature selects for or against.

Isolation is the separation of members of a species by environmental barriers that the organisms cannot cross. Environmental change, either gradual or sudden, often results in isolation. An example of gradual isolation is the formation of a mountain range or desert between members of a species. An example of sudden isolation is the separation of species members by a flood or earthquake. Isolation leads to evolution because the separated groups cannot reproduce together and differences arise. In addition, because the environment of each group is different, the groups adapt and evolve differently. Extended isolation can lead to speciation, the development of new species.

Sexual reproduction and selection contributes to evolution by consolidating genetic mutations and creating new combinations of genes. Genetic recombination during sexual reproduction introduces new combinations of traits and patterns of gene expression. Consolidation of favorable mutations through sexual reproduction speeds the processes of evolution and natural selection. On the other hand, consolidation of deleterious mutations creates completely unfit individuals that are readily eliminated from the population.

Genetic drift is, along with natural selection, one of the two main mechanisms of evolution. Genetic drift refers to the chance deviation in the frequency of alleles (traits) resulting from the randomness of zygote formation and selection. Because only a small percentage of all possible zygotes become mature adults, parents do not necessarily pass all of their alleles on to their offspring. Genetic drift is particularly important in small populations because chance deviations in allelic frequency can quickly alter the genotypic make-up of the population. In extreme cases, certain alleles may completely disappear from the gene pool. Genetic drift is particularly influential when environmental events and conditions produce small, isolated populations.

Plate tectonics is the theory that the Earth's surface consists of large plates. Movement and shifting of the plates dictate the location of continents, formation of mountains and seas, and volcanic and earthquake activity. Such contributions to environmental conditions influence the evolution of species. For example, tectonic activity resulting in mountain formation or continent separation can cause genetic isolation. In addition, the geographic distribution of species is indicative of evolutionary history and related tectonic activity.

Skill 8.5 Identify evidence for evolutionary change in organisms over time

The wide range of evidence of evolution provides information on the natural processes by which the variety of life on earth developed.

1. Palaeontology: Palaeontology is the study of past life based on fossil records and their relation to different geologic time periods. When organisms die, they often decompose quickly or are consumed by scavengers, leaving no evidence of their existence. Occasionally, some organisms are preserved. The remains or traces of the organisms from a past geological age embedded in rocks by natural processes are called fossils. They are very important for the understanding the evolutionary history of life on earth as they provide evidence of evolution and detailed information on the ancestry of organisms.

Petrification is the process by which a dead animal gets fossilized. For this to happen, a dead organism must be buried quickly to avoid weathering and decomposition. When the organism is buried, the organic matter decays. The mineral salts from the mud (in which the organism is buried) will infiltrate into the bones and gradually fill up the pores. The bones will harden and be preserved as fossils. If dead organisms are covered by wind- blown sand and if the sand is subsequently turned into mud by heavy rain or floods, the same process of mineral infiltration may occur. Besides petrification, the organisms may be well preserved in ice, in hardened resin of coniferous trees (amber), in tar or in anaerobic acidic peat. Fossilization can sometimes be a trace or an impression of a form – e.g., leaves and footprints.

Horizontal layers of sedimentary rocks (these are formed by silt or mud on top of each other) are called strata and each layer consists of fossils. The oldest layer is the one at the bottom of the pile and the fossils found in this layer are the oldest and this is how the paleaontologists determine the relative ages of these fossils.

Some organisms appear in some layers indicating that they lived only during that period and became extinct. A succession of animals and plants can also be seen in fossil records, which supports the theory that organisms tend to progressively increase in complexity.

According to fossil records, some modern species of plants and animals are found to be almost identical to the species that lived in ancient geological ages. They are existing species of ancient lineage that have remained unchanged morphologically and maybe physiologically as well. Hence they are called "living fossils". Some examples of living fossils are tuatara, nautilus, horseshoe crab, gingko and metasequoia.

2. Anatomy: Comparative anatomical studies reveal that some structural features are basically similar – e.g., flowers generally have sepals, petals, stigma, style and ovary but the size, color, number of petals or sepals may differ from species to species.

The degree of resemblance between two organisms indicates how closely they are related in evolution.

Groups with little in common are supposed to have diverged from a common ancestor much earlier in geological history than groups which have more in common.

To decide how closely two organisms are, anatomists look for the structures which may serve different purpose in the adult, but are basically similar in structure (homologous structures).

In cases where similar structures serve different functions in adults, it is important to trace their origin and embryonic development

When a group of organisms share a specialized homologous structure to perform a variety of functions it allows them to adapt to different environmental conditions. This is called adaptive radiation. The gradual spreading of organisms by adaptive radiation is known as divergent evolution. Examples of divergent evolution are pentadactyl limbs and insect mouthparts.

Under similar environmental conditions, fundamentally different structures in different groups of organisms may undergo modifications to serve similar functions. This is called convergent evolution. The structures, which have no close phylogenetic links but showing adaptation to perform the same functions, are called analogous structures. Examples include the wings of bats, bird and insects, the jointed legs of insects and vertebrates and the eyes of vertebrates and cephalopods.

Vestigial organs: Organs that are smaller and simpler in structure than corresponding parts in the ancestral species are called vestigial organs. They are usually degenerated or underdeveloped. These were functional in ancestral species but now have become nonfunctional, e.g., vestigial hind limbs of whales, vestigial leaves of some xerophytes or vestigial wings of flightless birds like ostriches.

3. Geographical distribution: may be continental or oceanic.

Continental distribution: All organisms are adapted to their environment to a greater or lesser extent. It is generally assumed that the same type of species would be found in a similar habitat in a similar geographic area.
Examples: Africa has short tailed (old world) monkeys, elephants, lions and giraffes. South America has long-tailed monkeys, pumas, jaguars and llamas
Evidence for migration and isolation: The fossil record shows that evolution of camels started in North America, from which they migrated across the Bering strait into Asia and Africa and through the Isthmus of Panama into south America.
Continental drift: Fossils of the ancient amphibians, arthropods and ferns are found in South America, Africa, India, Australia and Antarctica which can be dated to the Paleozoic Era, at which time they were all in a single landmass called Gondwana.
Oceanic Island distribution: Most small isolated islands only have native species.
Plant life in Hawaii could have arrived as airborne spores or as seeds in the droppings of birds. A few large mammals present in remote islands were brought by human settlers.

4. **Evidence from comparative embryology**: Comparative embryology shows how embryos start off looking the same. As they develop their similarities slowly decrease until they take the form of their particular class.
Example: Adult vertebrates are diverse, yet their embryos are quite similar at very early stages. Fishlike structures still form in early embryos of reptiles, birds and mammals. In fish embryos, a two-chambered heart, some veins, and parts of arteries develop and persist in adult fishes. The same structures form early in human embryos but do not persist in adults.

5. **Physiology and Biochemistry**:
Evolution of widely distributed proteins and molecules: All organisms make use of DNA and/or RNA. ATP is the metabolic currency. The genetic code is the same for every organism. A piece of RNA in a bacterial cell codes for the same protein as in a human cell.

Comparison of the DNA sequence allows organisms to be grouped by sequence similarity, and the resulting phylogenetic trees are typically consistent with traditional taxonomy and are often used to strengthen or correct taxonomic classifications. DNA sequence comparison is considered strong enough to be used to correct erroneous assumptions in the phylogenetic tree in cases where other evidence is missing. The sequence of the 168rRNA gene, a vital gene encoding a part of the ribosome, was used to find the broad phylogenetic relationships between all life.

Proteomic evidence also supports the universal ancestry of life. Vital proteins such as ribosomes, DNA polymerase, and RNA polymerase are found in the most primitive bacteria to the most complex mammals. Since metabolic processes do not leave fossils, research into the evolution of the basic cellular processes is done largely by comparison of existing organisms.

COMPETENCE 9.0 UNDERSTAND HUMAN BIOLOGY, INCLUDING THE INTERACTIONS OF HUMANS WITH THE ENVIRONMENT

Skill 9.1 Demonstrate knowledge of human anatomy and physiology and how humans meet their basic needs

Skeletal System - The skeletal system functions in support. Vertebrates have an endoskeleton, with muscles attached to bones. Skeletal proportions are controlled by area to volume relationships. Body size and shape are limited due to the forces of gravity. Surface area is increased to improve efficiency in all organ systems.

The **axial skeleton** consists of the bones of the skull and vertebrae. The **appendicular skeleton** consists of the bones of the legs, arms, tail, and shoulder girdle. Bone is a connective tissue. Parts of the bone include compact bone which gives strength, spongy bone which contains red marrow to make blood cells, yellow marrow in the center of long bones to store fat cells, and the periosteum which is the protective covering on the outside of the bone.

A **joint** is defined as a place where two bones meet. Joints enable movement. **Ligaments** attach bone to bone. **Tendons** attach bones to muscles.

Muscular System - Functions in movement. There are three types of muscle tissue. Skeletal muscle is voluntary. These muscles are attached to bones. Smooth muscle is involuntary. It is found in organs and enables functions such as digestion and respiration. Cardiac muscle is a specialized type of smooth muscle and is found in the heart. Muscles can only contract; therefore they work in antagonistic pairs to allow back and forward movement. Muscle fibers are made of groups of myofibrils which are made of groups of sarcomeres. Actin and myosin are proteins which make up the sarcomere.

Physiology of muscle contraction - A nerve impulse strikes a muscle fiber. This causes calcium ions to flood the sarcomere. Calcium ions allow ATP to expend energy. The myosin fibers creep along the actin, causing the muscle to contract. Once the nerve impulse has passed, calcium is pumped out and the contraction ends.

Nervous System - The neuron is the basic unit of the nervous system. It consists of an axon, which carries impulses away from the cell body, the dendrite, which carries impulses toward the cell body, and the cell body, which contains the nucleus. Synapses are spaces between neurons. Chemicals called neurotransmitters are found close to the synapse. The myelin sheath, composed of Schwann cells, covers the neurons and provides insulation.

Physiology of the nerve impulse - Nerve action depends on depolarization and an imbalance of electrical charges across the neuron. A polarized nerve has a positive charge outside the neuron. A depolarized nerve has a negative charge outside the neuron. Neurotransmitters turn off the sodium pump which results in depolarization of the membrane. This wave of depolarization (as it moves from neuron to neuron) carries an electrical impulse. This is actually a wave of opening and closing gates that allows for the flow of ions across the synapse. Nerves have an action potential. There is a threshold of the level of chemicals that must be met or exceeded in order for muscles to respond. This is called the "all or none" response.

The **reflex arc** is the simplest nerve response. The brain is bypassed. When a stimulus (like touching a hot stove) occurs, sensors in the hand send the message directly to the spinal cord. This stimulates motor neurons that contract the muscles to move the hand.

Voluntary nerve responses involve the brain. Receptor cells send the message to sensory neurons which lead to association neurons. The message is taken to the brain. Motor neurons are stimulated and the message is transmitted to effector cells which cause the end effect.

Organization of the Nervous System - The somatic nervous system is controlled consciously. It consists of the central nervous system (brain and spinal cord) and the peripheral nervous system (nerves that extend from the spinal cord to the muscles). The autonomic nervous system is unconsciously controlled by the hypothalamus of the brain. Smooth muscles, the heart and digestion are some processes controlled by the autonomic nervous system. The sympathetic nervous system works opposite of the parasympathetic nervous system. For example, if the sympathetic nervous system stimulates an action, the parasympathetic nervous system would end that action.

Neurotransmitters - these are chemicals released by exocytosis. Some neurotransmitters stimulate, while others inhibit, action.

Acetylcholine - the most common neurotransmitter; it controls muscle contraction and heartbeat. The enzyme acetylcholinesterase breaks it down to end the transmission.

Epinephrine - responsible for the "fight or flight" reaction. It causes an increase in heart rate and blood flow to prepare the body for action. It is also called adrenaline.

Endorphins and enkephalins - these are natural pain killers and are released during serious injury and childbirth.

Digestive System - The function of the digestive system is to break food down and absorb it into the blood stream where it can be delivered to all cells of the body for use in cellular respiration. The teeth and saliva begin digestion by breaking food down into smaller pieces and lubricating it so it can be swallowed. The lips, cheeks, and tongue form a bolus (ball) of food. It is carried down the pharynx by the process of peristalsis (wave like contractions) and enters the stomach through the cardiac sphincter which closes to keep food from going back up. In the stomach, pepsinogen and hydrochloric acid form pepsin, the enzyme that breaks down proteins. The food is broken down further by this chemical action and is turned into chyme. The pyloric sphincter muscle opens to allow the food to enter the small intestine. Most nutrient absorption occurs in the small intestine. Its large surface area, accomplished by its length and protrusions called villi and microvilli allow for a great absorptive surface. Upon arrival into the small intestine, chyme is neutralized to allow the enzymes found there to function. Any food left after the trip through the small intestine enters the large intestine. The large intestine functions to reabsorb water and produce vitamin K. The feces, or remaining waste, are passed out through the anus.

Accessory organs - although not part of the digestive tract, these organs function in the production of necessary enzymes and bile. The pancreas makes many enzymes to break down food in the small intestine. The liver makes bile which breaks down and emulsifies fatty acids.

Respiratory System - This system functions in the gas exchange of oxygen (needed) and carbon dioxide (waste). It delivers oxygen to the bloodstream and picks up carbon dioxide for release out of the body. Air enters the mouth and nose, where it is warmed, moistened and filtered of dust and particles. Cilia in the trachea trap unwanted material in mucus, which can be expelled. The trachea splits into two bronchial tubes and the bronchial tubes divide into smaller and smaller bronchioles in the lungs. The internal surface of the lung is composed of alveoli, which are thin walled air sacs. These allow for a large surface area for gas exchange. The alveoli are lined with capillaries. Oxygen diffuses into the bloodstream and carbon dioxide diffuses out to be exhaled. The oxygenated blood is carried to the heart and delivered to all parts of the body.

The thoracic cavity holds the lungs. A muscle, the diaphragm, below the lungs is an adaptation that makes inhalation possible. As the volume of the thoracic cavity increases, the diaphragm muscle flattens out and inhalation occurs. When the diaphragm relaxes, exhalation occurs.

Circulatory System - The function of the circulatory system is to carry oxygenated blood and nutrients to all cells of the body and return carbon dioxide waste to be expelled from the lungs. Be familiar with the parts of the heart and the path blood takes from the heart to the lungs, through the body and back to the heart. Unoxygenated blood enters the heart through the inferior and superior vena cava. The first chamber it encounters is the right atrium. It goes through the tricuspid valve to the right ventricle, on to the pulmonary arteries, and then to the lungs where it is oxygenated. It returns to the heart through the pulmonary vein into the left atrium. It travels through the bicuspid valve to the left ventricle where it is pumped to all parts of the body through the aorta.

Sinoatrial node (SA node) - the pacemaker of the heart. Located on the right atrium, it is responsible for contraction of the right and left atrium.

Atrioventricular node (AV node) - located on the left ventricle, it is responsible for contraction of the ventricles.

Blood vessels include:

- **arteries** - lead away from the heart. All arteries carry oxygenated blood except the pulmonary artery going to the lungs. Arteries are under high pressure.
- **arterioles** - arteries branch off to form these smaller passages.
- **capillaries** - arterioles branch off to form tiny capillaries that reach every cell. Blood moves slowest here due to the small size; only one red blood cell may pass at a time to allow for diffusion of gases into and out of cells. Nutrients are also absorbed by the cells from the capillaries.
- **venules** - capillaries combine to form larger venules. The vessels are now carrying waste products from the cells.
- **veins** - venules combine to form larger veins, leading back to the heart. Veins and venules have thinner walls than arteries because they are not under as much pressure. Veins contain valves to prevent the backward flow of blood due to gravity.

Components of the blood include:

- **plasma** – 60% of the blood is plasma. It contains salts called electrolytes, nutrients, and waste. It is the liquid part of blood.
- **erythrocytes** - also called red blood cells; they contain hemoglobin which carries oxygen molecules.
- **leukocytes** - also called white blood cells. White blood cells are larger than red cells. They are phagocytic and can engulf invaders. White blood cells are not confined to the blood vessels and can enter the interstitial fluid between cells.
- **platelets** - assist in blood clotting. Platelets are made in the bone marrow.
- **Blood clotting** - the neurotransmitter that initiates blood vessel constriction following an injury is called serotonin. A material called prothrombin is converted to thrombin with the help of thromboplastin. The thrombin is then used to convert fibrinogen to fibrin which traps red blood cells to form a scab and stop blood flow.

Lymphatic System (Immune System)

Nonspecific defense mechanisms – They do not target specific pathogens, but are a whole body response. Results of nonspecific mechanisms are seen as symptoms of an infection. These mechanisms include the skin, mucous membranes and cells of the blood and lymph (ie: white blood cells, macrophages). Fever is a result of an increase of white blood cells. Pyrogens are released by white blood cells which set the body's thermostat to a higher temperature. This inhibits the growth of microorganisms. It also increases metabolism to increase phagocytosis and body repair.

Specific defense mechanisms - They recognize foreign material and respond by destroying the invader. These mechanisms are specific in purpose and diverse in type. They are able to recognize individual pathogens. They are able to differntiate between foreign material and self. Memory of the invaders provides immunity upon further exposure.

antigen - any foreign particle that invades the body.
antibody - manufactured by the body, they recognize and latch onto antigens, hopefully destroying them.
immunity - this is the body's ability to recognize and destroy an antigen before it causes harm. Active immunity develops after recovery from an infectious disease (chicken pox) or after a vaccination (mumps, measles, rubella). Passive immunity may be passed from one individual to another. It is not permanent. A good example is the immunities passed from mother to nursing child.

Excretory System
The function of the excretory system is to rid the body of nitrogenous wastes in the form of urea. The functional unit of excretion are the nephrons, which make up the kidneys. Antidiuretic hormone (ADH), which is made in the hypothalamus and stored in the pituitary, is released when differences in osmotic balance occur. This will cause more water to be reabsorbed. As the blood becomes more dilute, ADH release ceases.

The Bowman's capsule contains the glomerulus, a tightly packed group of capillaries. The glomerulus is under high pressure. Waste and fluids leak out due to pressure. Filtration is not selective in this area. Selective secretion by active and passive transport occur in the proximal convoluted tubule. Unwanted molecules are secreted into the filtrate. Selective secretion also occurs in the loop of Henle. Salt is actively pumped out of the tube and much water is lost due to the hyperosmosity of the inner part (medulla) of the kidney. As the fluid enters the distal convoluted tubule, more water is reabsorbed. Urine forms in the collecting duct which leads to the ureter then to the bladder where it is stored. Urine is passed from the bladder through the urethra. The amount of water reabsorbed back into the body is dependent upon how much water or fluids an individual has consumed. Urine can be very dilute or very concentrated if dehydration is present.

Endocrine System

The function of the endocrine system is to manufacture proteins called hormones. Hormones are released into the bloodstream and are carried to a target tissue where they stimulate an action. Hormones may build up over time to cause their effect, as in puberty or the menstrual cycle.

Hormone activation - Hormones are specific and fit receptors on the target tissue cell surface. The receptor activates an enzyme which converts ATP to cyclic AMP. Cyclic AMP (cAMP) is a second messenger from the cell membrane to the nucleus. The genes found in the nucleus turn on or off to cause a specific response.

There are two classes of hormones. **Steroid hormones** come from cholesterol. Steroid hormones cause sexual characteristics and mating behavior. Hormones include estrogen and progesterone in females and testosterone in males. **Peptide hormones** are made in the pituitary, adrenal glands (kidneys), and the pancreas. They include the following:

- **Follicle stimulating hormone (FSH)** - production of sperm or egg cells
- **Luteinizing hormone (LH)** - functions in ovulation
- **Luteotropic hormone (LTH)** - assists in production of progesterone
- **Growth hormone (GH)** - stimulates growth
- **Antidiuretic hormone (ADH)** - assists in retention of water

- **Oxytocin** - stimulates labor contractions at birth and let-down of milk
- **Melatonin** - regulates circadian rhythms and seasonal changes
- **Epinephrine (adrenaline)** - causes fight or flight reaction of the nervous system
- **Thyroxin** - increases metabolic rate
- **Calcitonin** - removes calcium from the blood
- **Insulin** - decreases glucose level in blood
- **Glucagon** - increases glucose level in blood

Hormones work on a feedback system. The increase or decrease in one hormone may cause the increase or decrease in another. Release of hormones cause a specific response.

Reproductive System

Sexual reproduction greatly increases diversity due to the many combinations possible through meiosis and fertilization. Gametogenesis is the production of the sperm and egg cells. Spermatogenesis begins at puberty in the male. One spermatozoa produces four sperm. The sperm mature in the seminiferous tubules located in the testes. Oogenesis, the production of egg cells is usually complete by the birth of a female. Egg cells are not released until menstruation begins at puberty. Meiosis forms one ovum with all the cytoplasm and three polar bodies which are reabsorbed by the body. The ovum are stored in the ovaries and released each month from puberty to menopause.

Path of the sperm - sperm are stored in the seminiferous tubules in the testes where they mature. Mature sperm are found in the epididymis located on top of the testes. After ejaculation, the sperm travels up the vas deferens where they mix with semen made in the prostate and seminal vesicles and travel out the urethra.

Path of the egg - eggs are stored in the ovaries. Ovulation releases the egg into the fallopian tubes which are ciliated to move the egg along. Fertilization normally occurs in the fallopian tube. If pregnancy does not occur, the egg passes through the uterus and is expelled through the vagina during menstruation. Levels of progesterone and estrogen stimulate menstruation. In the event of pregnancy, hormonal levels are affected by the implantation of a fertilized egg, so menstruation does not occur.

Pregnancy - if fertilization occurs, the zygote implants in about two to three days in the uterus. Implantation promotes secretion of human chorionic gonadotropin (HCG). This is what is detected in pregnancy tests. The HCG keeps the level of progesterone elevated to maintain the uterine lining in order to feed the developing embryo until the umbilical cord forms. Labor is initiated by oxytocin which causes labor contractions and dilation of the cervix. Prolactin and oxytocin cause the production of milk.

Skill 9.2 Identify responsible health practices, including issues relating to nutrition and exercise

Good nutrition is paramount in maintaining health for growth and development. A balanced diet includes foods from the major food groups of carbohydrates, proteins, lipids, and sufficient quantities of vitamins and minerals.

Body pollutants, such as tobacco, drugs, and alcohol interfere with the absorption of nutrients and also may interfere with physical and mental development. They may also damage developing organs, leading to lifelong diseases such as emphysema or asthma.

Skill 9.3 Analyze the global impact humans have on living and nonliving environments, including issues related to overpopulation

Ecological Problems - nonrenewable resources are fragile and must be conserved for use in the future. Man's impact and knowledge of conservation will control our future.

Biological magnification - chemicals and pesticides accumulate along the food chain. Tertiary consumers have more accumulated toxins than animals at the bottom of the food chain.

Simplification of the food web - Three major crops feed the world (rice, corn, wheat). The planting of these foods wipe out habitats and push animals residing there into other habitats causing overpopulation or extinction.

Fuel sources - strip mining and the overuse of oil reserves have depleted these resources. At the current rate of consumption, conservation or alternate fuel sources will guarantee our future fuel sources.

Pollution - although technology gives us many advances, pollution is a side effect of production. Waste disposal and the burning of fossil fuels have polluted our land, water and air. Global warming and acid rain are two results of the burning of hydrocarbons and sulfur.

Global warming - rainforest depletion and the use of fossil fuels and aerosols have caused an increase in carbon dioxide production. This leads to a decrease in the amount of oxygen which is directly proportional to the amount of ozone. As the ozone layer depletes, more heat enters our atmosphere and is trapped. This causes an overall warming effect which may eventually melt polar ice caps, causing a rise in water levels and changes in climate which will affect weather systems worldwide.

Endangered species - construction of homes to house people in our overpopulated world has caused the destruction of habitat for other animals leading to their extinction.

Overpopulation - the human race is still growing at an exponential rate. Carrying capacity has not been met due to our ability to use technology to produce more food and housing. Space and water can not be manufactured and eventually our non-renewable resources will reach a crisis state. Our overuse affects every living thing on this planet.

Pollutants are impurities in air and water that may be harmful to life. Spills from barges carrying large quantities of oil pollute beaches and harm fish.

All acids contain hydrogen. Acidic substances from factories and car exhausts dissolve in rainwater forming **acid rain.** Acid rain forms predominantly from pollutant oxides in the air (usually nitrogen-based NO_x or sulfur-based SO_x), which become hydrated into their acids (nitric or sulfuric acid). When the rain falls into stone, the acids can react with metallic compounds and gradually wear the stone away.

Groundwater provides drinking water for 53% of the population in the United States. Much groundwater is clean enough to drink without any type of treatment. Impurities in the water are filtered out by the rocks and soil through which it flows. However, many groundwater sources are becoming contaminated. Septic tanks, broken pipes, agriculture fertilizers, garbage dumps, rainwater runoff and leaking underground tanks all pollute groundwater. Toxic chemicals from farmland mix with groundwater. Removal of large volumes of groundwater can cause collapse of soil and rock underground, causing the ground to sink. Along shorelines, excessive depletion of underground water supplies allows the intrusion of saltwater into the freshwater field. The groundwater supply becomes undrinkable.

Skill 9.4 Analyze the relationships among renewable and nonrenewable resources and the human population

A **renewable resource** is one that is replaced naturally. Living renewable resources would be plants and animals. Plants are renewable because they grow and reproduce. Sometimes renewal of the resource doesn't keep up with the demand. Such is the case with trees. Since the housing industry uses lumber for frames and homebuilding they are often cut down faster than new trees can grow. Now there are specific tree farms. Special methods allow trees to grow faster.

A second renewable resource is animals. They renew by the process of reproduction. Some wild animals need protection on refuges. As the population of humans increases resources are used faster. Cattle are used for their hides and for food. Some animals like deer are killed for sport. Each state has an environmental protection agency with divisions of forest management and wildlife management.

Non-living renewable resources would be water, air, and soil. Water is renewed in a natural cycle called the water cycle. Air is a mixture of gases. Oxygen is given off by plants and taken in by animals that in turn expel the carbon dioxide that the plants need. Soil is another renewable resource. Fertile soil is rich in minerals. When plants grow they remove the minerals and make the soil less fertile. Chemical treatments are one way or renewing the composition. It is also accomplished naturally when the plants decay back into the soil. The plant material is used to make compost to mix with the soil.

Nonrenewable resources are not easily replaced in a timely fashion. Minerals are nonrenewable resources. Quartz, mica, salt and sulfur are some examples. Mining depletes these resources so society may benefit. Glass is made from quartz, electronic equipment from mica, and salt has many uses. Sulfur is used in medicine, fertilizers, paper and matches.

Metals are among the most widely used nonrenewable resource. Metals must be separated from the ore. Iron is our most important ore. Gold, silver and copper are often found in a more pure form called native metals.

COMPETENCY 10.0 UNDERSTAND INTERACTIONS OF ORGANISMS WITH ONE ANOTHER AND THEIR ENVIRONMENTS

Skill 10.1 Identify characteristics of populations, communities, ecosystems, and biomes

Ecology is the study of organisms, where they live and their interactions with the environment. An ecosystem is a community along with its environment, all functioning together. A **population** is a group of the same species in a specific area. A **community** is a group of populations residing in the same area. Communities that are ecologically similar in regards to temperature, rainfall and the species that live there are called **biomes**. Specific biomes include:

Marine - covers 75% of the earth. This biome is organized by the depth of the water. The intertidal zone is from the tide line to the edge of the water. The littoral zone is from the water's edge to the open sea. It includes coral reef habitats and is the most densely populated area of the marine biome. The open sea zone is divided into the epipelagic zone and the pelagic zone. The epipelagic zone receives more sunlight and has a larger number of species. The ocean floor is called the benthic zone and is populated with bottom feeders.

Tropical Rain Forest - temperature is constant (25 degrees C), rainfall exceeds 200 cm. per year. Located around the area of the equator, the rain forest has abundant, diverse species of plants and animals.

Savanna - temperatures range from 0-25 degrees C depending on the location. Rainfall is from 90 to 150 cm per year. Plants include shrubs and grasses. The savanna is a transitional biome between the rain forest and the desert.

Desert - temperatures range from 10-38 degrees C. Rainfall is under 25 cm per year. Plant species include xerophytes and succulents. Lizards, snakes and small mammals are common animals.

Temperate Deciduous Forest - temperature ranges from -24 to 38 degrees C. Rainfall is between 65 to 150 cm per year. Deciduous trees are common, as well as deer, bear and squirrels.

Taiga - temperatures range from -24 to 22 degrees C. Rainfall is between 35 to 40 cm per year. Taiga is located very north and very south of the equator, getting close to the poles. Plant life includes conifers and plants that can withstand harsh winters. Animals include weasels, mink, and moose.

Tundra - temperatures range from -28 to 15 degrees C. Rainfall is limited, ranging from 10 to 15 cm per year. The tundra is located even further north and south than the taiga. Common plants include lichens and mosses. Animals include polar bears and musk ox.

Polar or Permafrost - temperature ranges from -40 to 0 degrees C. It rarely gets above freezing. Rainfall is below 10 cm per year. Most water is bound up as ice. Life is limited.

Succession - Succession is an orderly process of replacing a community that has been damaged or beginning one where no life previously existed. Primary succession occurs after a community has been totally wiped out by a natural disaster or where life never existed before, as in a flooded area. Secondary succession takes place in communities that were once flourishing but were disturbed by some source, either man or nature, but were not totally stripped. A climax community is a community that is established and flourishing.

Skill 10.2 Demonstrate an understanding of the interactions of living and nonliving components of ecosystems and of limiting factors that regulate productivity, complexity, and population sizes within ecosystems (e.g., temperature, soil fertility, light intensity)

Biotic factors - living things in an ecosystem; plants, animals, bacteria, fungi, etc. If one population in a community increases, it affects the ability of another population to succeed by limiting the available amount of food, water, shelter and space.

Abiotic factors - nonliving aspects of an ecosystem; soil quality, rainfall, and temperature. Changes in climate and soil can cause effects at the beginning of the food chain, thus limiting or accelerating the growth of populations.

A **limiting factor** is the component of a biological process that determines how quickly or slowly the process proceeds. Photosynthesis is the main biological process determining the rate of ecosystem productivity, the rate at which an ecosystem creates biomass. Thus, in evaluating the productivity of an ecosystem, potential limiting factors are light intensity, gas concentrations, and mineral availability. The Law of the Minimum states that the required factor in a given process that is most scarce controls the rate of the process.

One potential limiting factor of ecosystem productivity is light intensity because photosynthesis requires light energy. Light intensity can limit productivity in two ways. First, too little light limits the rate of photosynthesis because the required energy is not available. Second, too much light can damage the photosynthetic system of plants and microorganisms thus slowing the rate of photosynthesis. Decreased photosynthesis equals decreased productivity.

Another potential limiting factor of ecosystem productivity is gas concentrations. Photosynthesis requires carbon dioxide. Thus, increased concentration of carbon dioxide often results in increased productivity. While carbon dioxide is often not the ultimate limiting factor of productivity, increased concentration can indirectly increase rates of photosynthesis in several ways. First, increased carbon dioxide concentration often increases the rate of nitrogen fixation (available nitrogen is another limiting factor of productivity). Second, increased carbon dioxide concentration can decrease the pH of rain, improving the water source of photosynthetic organisms that live in basic soils.

Finally, mineral availability also limits ecosystem productivity. Plants require adequate amounts of nitrogen and phosphorus to build many cellular structures. The availability of the inorganic minerals phosphorus and nitrogen often is the main limiting factor of plant biomass production. In other words, in a natural environment phosphorus and nitrogen availability most often limits ecosystem productivity, rather than carbon dioxide concentration or light intensity.

Skill 10.3 Demonstrate an understanding of the concepts of niche and carrying capacity

The term 'Niche' describes the relational position of a species or population in an ecosystem. Niche includes how a population responds to the abundance of its resources and enemies (e.g., by growing when resources are abundant and predators, parasites and pathogens are scarce).

Niche also indicates the life history of an organism, habitat and place in the food chain.

According to the competitive exclusion principle, no two species can occupy the same niche in the same environment for a long time.

The full range of environmental conditions (biological and physical) under which an organism can exist describes its fundamental niche. Because of the pressure from superior competitors, inferiors are driven to occupy a niche much narrower than their previous niche. This is known as the 'realized niche'

Examples of niche:

1. Oak trees:

* live in forests
* sunlight is used for photosynthesis
* provide shelter for many animals
* act as support for creeping plants
* serve as a source of food for animals
* cover their ground with dead leaves in the autumn

If the oak trees were cut down or destroyed by fire or storms they would no longer be doing their job and this would have a disastrous effect on all the other organisms living in the same habitat.

2. Hedgehogs:

* eat a variety of insects and other invertebrates which live underneath the dead leaves and twigs in the garden
* the spines are a superb environment for fleas and ticks
* put the nitrogen back into the soil when they urinate
* eat slugs and protect plants from them

If there were no hedgehogs around, the population of slugs would increase and the nutrients in the dead leaves and twigs would not be recycled.

A **population** is a group of individuals of one species that live in the same general area. Many factors can affect the population size and its growth rate. Population size can depend on the total amount of life a habitat can support. This is the carrying capacity of the environment. Once the habitat runs out of food, water, shelter, or space, the carrying capacity decreases, and then stabilizes.

Limiting factors can affect population growth. As a population increases, the competition for resources is more intense, and the growth rate declines. This is a **density-dependent** growth factor. The carrying capacity can be determined by the density-dependent factor. **Density-independent factors** affect the individuals regardless of population size. The weather and climate are good examples. Too hot or too cold temperatures may kill many individuals from a population that has not reached its carrying capacity.

Human population increased slowly until 1650. Since 1650, the human population has grown almost exponentially, reaching its current population of over 6 billion. Factors that have led to this increased growth rate include improved nutrition, sanitation and health care. In addition, advances in technology, agriculture and scientific knowledge have made the use of resources more efficient and increased their availability.

Carrying Capacity - this is the total amount of life a habitat can support. Once the habitat runs out of food, water, shelter, or space, the carrying capacity decreases, and then stabilizes.

While the Earth's ultimate carrying capacity for humans is uncertain, some factors that may limit growth are the availability of food, water, space, and fossil fuels. There is a finite amount of land on Earth available for food production. In addition, providing clean, potable water for a growing human population is a real concern. Finally, fossil fuels, important energy sources for human technology, are scarce. The inevitable shortage of energy in the Earth's future will require the development of alternative energy sources to maintain or increase human population growth.

Skill 10.4 Analyze the roles of organisms within an ecosystem and the interrelationships among organisms in ecosystems (e.g., predator-prey, commensalism, parasitism)

Definitions of feeding relationships:

Parasitism - two species that occupy a similar place; the parasite benefits from the relationship, the host is harmed.

Commensalism - two species that occupy a similar place; neither species is harmed or benefits from the relationship.

Mutualism (symbiosis)- two species that occupy a similar place; both species benefit from the relationship.

Competition - two species that occupy the same habitat or eat the same food are said to be in competition with each other.

Predation - animals that eat other animals are called predators. The animals they feed on are called the prey. Population growth depends upon competition for food, water, shelter, and space. The amount of predators determines the amount of prey, which in turn affects the number of predators.

Skill 10.5 Analyze the flow of energy and the features of food chains and food webs in various types of ecosystems

Trophic levels are based on the feeding relationships that determine energy flow and chemical cycling.

Autotrophs are the primary producers of the ecosystem. **Producers** mainly consist of plants. **Primary consumers** are the next trophic level. The primary consumers are the herbivores that eat plants or algae. **Secondary consumers** are the carnivores that eat the primary consumers. **Tertiary consumers** eat the secondary consumer. These trophic levels may go higher depending on the ecosystem. **Decomposers** are consumers that feed off animal waste and dead organisms. This pathway of food transfer is known as the food chain.

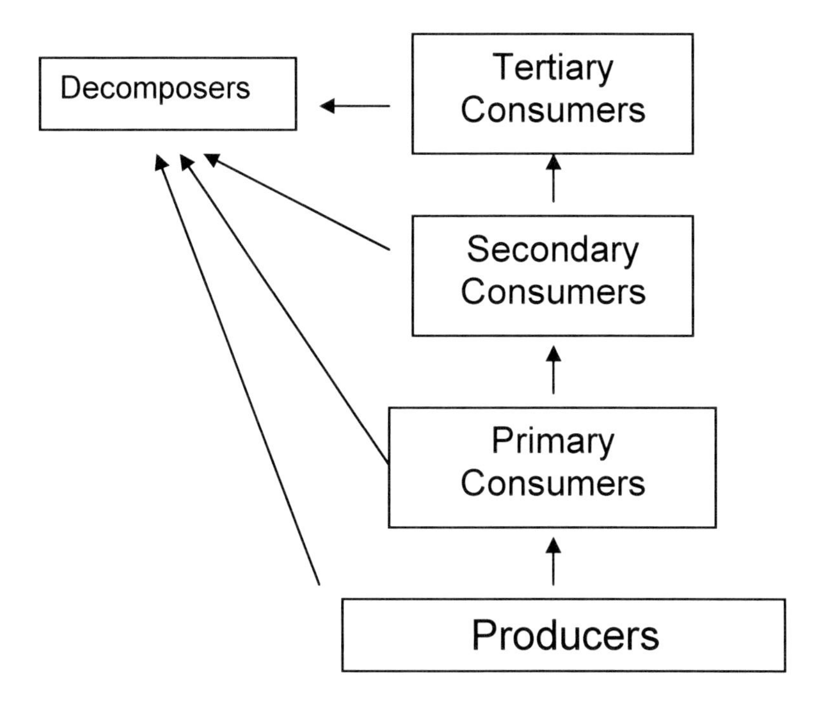

Most food chains are more elaborate, becoming food webs.

Energy is lost as the trophic levels progress from producer to tertiary consumer. The amount of energy that is transferred between trophic levels is called the ecological efficiency. The visual of this energy flow is represented in a **pyramid of productivity**.

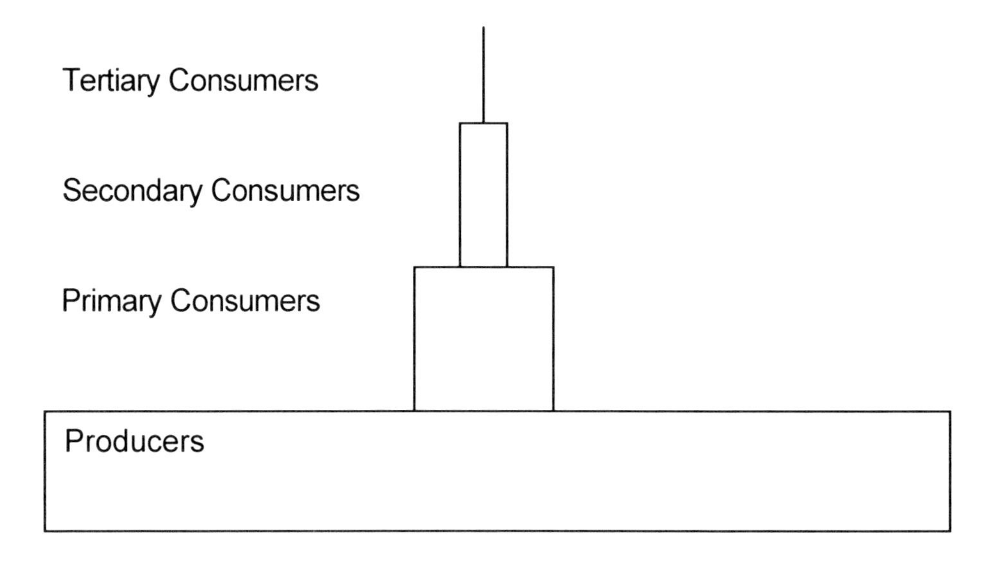

The **biomass pyramid** represents the total dry weight of organisms in each trophic level. A **pyramid of numbers** is a representation of the population size of each trophic level. The producers, being the most populous, are on the bottom of this pyramid with the tertiary consumers on the top with the fewest numbers.

SUBAREA III. PHYSICAL SCIENCE

COMPETENCE 11.0 UNDERSTAND THE PROPERTIES OF MATTER

Skill 11.1 Demonstrate knowledge of the atomic and subatomic structure of matter

The atomic theory of matter suggests that:

1. All matter consists of atoms
2. All atoms of an element are identical
3. Different elements have different atoms
4. Atoms maintain their properties in a chemical reaction

The Danish scientist Neils Bohr created a model in 1913. The results of his model are:

1. Electrons orbit the nucleus, but only certain orbits are allowed. An electron in an allowed orbit will not lose energy.
2. When an electron moves from an outer orbit to an inner orbit, it gives off energy.
3. When an electron moves from an inner orbit to an outer orbit, it absorbs energy.

Bohr's model only explains the very simplest atoms, such as hydrogen. Today's more sophisticated atomic model is based upon how waves react.

An atom is a nucleus surrounded by a cloud with moving electrons.

The nucleus is the center of the atom. The positive particles inside the nucleus are called protons. The mass of a proton is about 2,000 times that of the mass of an electron. The number of protons in the nucleus of an atom is called the atomic number. All atoms of the same element have the same atomic number.

Neutrons are another type of particle in the nucleus. Neutrons and protons have about the same mass, but neutrons have no charge. Neutrons were discovered because scientists observed that not all atoms in neon gas have the same mass. They had identified isotopes. Isotopes of an element have the same number of protons in the nucleus, but have different masses. Neutrons explain the difference in mass. They have mass but no charge.

The mass of matter is measured against a standard mass such as the gram. Scientists measure the mass of an atom by comparing it to that of a standard atom. The result is relative mass. The relative mass of an atom is its mass expressed in terms of the mass of the standard atom. The isotope of the element carbon is the standard atom. It has six (6) neutrons and is called carbon-12.It is assigned a mass of 12 atomic mass units (amu). Therefore, the atomic mass unit (amu) is the standard unit for measuring the mass of an atom. It is equal to the mass of a carbon atom.

The mass number of an atom is the sum of its protons and neutrons. In any element, there is a mixture of isotopes, some having slightly more or slightly fewer protons and neutrons. The atomic mass of an element is an average of the mass numbers of its atoms.

The following table summarizes the terms used to describe atomic nuclei:

Term	Meaning	Characteristic	
Atomic Number	# protons (p) in a given element	same for all	Carbon (C) atomic number = 6 (6p)
Mass number	# protons + # neutrons (p + n)	changes for different isotopes of an element	C-12 (6p + 6n) C-13 (6p + 7n)
Atomic mass (average mass)	mass of an atom	atomic mass of carbon equals 12.011	

Each atom has an equal number of electrons (negative) and protons (positive). Therefore, atoms are neutral. Electrons orbiting the nucleus occupy energy levels that are arranged in order and the electrons tend to occupy the lowest energy level available. A stable electron arrangement is an atom that has all of its electrons in the lowest possible energy levels.

Each energy level holds a maximum number of electrons. However, an atom with more than one level does not hold more than 8 electrons in its outermost shell.

Level	Name	Max. # of Electrons
First	K shell	2
Second	L shell	8
Third	M shell	18
Fourth	N shell	32

This can help explain why chemical reactions occur. Atoms react with each other when their outer levels are unfilled. When atoms either exchange or share electrons with each other, these energy levels become filled and the atom becomes more stable.

As an electron gains energy, it moves from one energy level to a higher energy level. The electron can not leave one level until it has enough energy to reach the next level. Excited electrons are electrons that have absorbed energy and have moved farther from the nucleus.

Electrons can also lose energy. When they do, they fall to a lower level. However, they can only fall to the lowest level that has room for them. This explains why atoms do not collapse.

Skill 11.2 Compare and contrast the characteristics of atoms, elements, molecules, and compounds

An element is a substance that can not be broken down into other substances. To date, scientists have identified 109 elements: 89 are found in nature and 20 are synthetic.

An atom is the smallest particle of the element that retains the properties of that element. All of the atoms of a particular element are the same. The atoms of each element are different from the atoms of other elements.

Elements are assigned an identifying symbol of one or two letters. The symbol for oxygen is O and stands for one atom of oxygen. However, because oxygen atoms in nature are joined together is pairs, the symbol O_2 represents oxygen. This pair of oxygen atoms is a molecule. A molecule is the smallest particle of substance that can exist independently and has all of the properties of that substance. A molecule of most elements is made up of one atom. However, oxygen, hydrogen, nitrogen, and chlorine molecules are made of two atoms each.

A compound is made of two or more elements that have been chemically combined. Atoms join together when elements are chemically combined. The result is that the elements lose their individual identities when they are joined. The compound that they become has different properties.

We use a formula to show the elements of a chemical compound. A chemical formula is a shorthand way of showing what is in a compound by using symbols and subscripts. The letter symbols let us know what elements are involved and the number subscript tells how many atoms of each element are involved. No subscript is used if there is only one atom involved. For example, carbon dioxide is made up of one atom of carbon (C) and two atoms of oxygen (O_2), so the formula would be represented as CO_2.

Substances can combine without a chemical change. A mixture is any combination of two or more substances in which the substances keep their own properties. A fruit salad is a mixture. So is an ice cream sundae, although you might not recognize each part if it is stirred together. Colognes and perfumes are the other examples. You may not readily recognize the individual elements. However, they can be separated.

Compounds and mixtures are similar in that they are made up of two or more substances. However, they have the following opposite characteristics:

Compounds:

1. Made up of one kind of particle
2. Formed during a chemical change
3. Broken down only by chemical changes
4. Properties are different from its parts
5. Has a specific amount of each ingredient.

Mixtures:

1. Made up of two or more particles
2. Not formed by a chemical change
3. Can be separated by physical changes
4. Properties are the same as its parts.
5. Does not have a definite amount of each ingredient.

Common compounds are acids, bases, salts, and oxides and are classified according to their characteristics.

Skill 11.3 Identify the physical and chemical properties of elements and compounds (e.g., density, boiling point, solubility)

DENSITY:

Everything in our world is made up of matter, whether it is a rock, a building, an animal, or a person. Matter is defined by its characteristics: It takes up space and it has mass.

Density is the mass of a substance contained per unit of volume. If the density of an object is less than the density of a liquid, the object will float in the liquid. If the object is denser than the liquid, then the object will sink.

Density is stated in grams per cubic centimeter (g/cm^3) where the gram is the standard unit of mass. To find an object's density, you must measure its mass and its volume. Then divide the mass by the volume ($D = m/V$).

To discover an object's density, first use a balance to find its mass. Then calculate its volume. If the object is a regular shape, you can find the volume by multiplying the length, width, and height together. However, if it is an irregular shape, you can find the volume by seeing how much water it displaces. Measure the water in the container before and after the object is submerged. The difference will be the volume of the object.

BOILING POINT:

Boiling point refers to the temperature at which a liquid becomes a gas. Boiling occurs when there is enough energy available to break the intermolecular forces holding molecules together as a liquid.

SOLUBILITY:

A material's **solubility** refers to the amount of solute (substance which will be dissolved) that will dissolve in a specific fluid (solvent) under given conditions. The solvent and solute together are known as the solution, and the process of dissolving is called solvation.

Skill 11.4 Interpret chemical symbols, formulas, and expressions

One or more substances are formed during a **chemical reaction**. Also, energy is released during some chemical reactions. Sometimes the energy release is slow and sometimes it is rapid. In a fireworks display, energy is released very rapidly. However, the chemical reaction that produces tarnish on a silver spoon happens very slowly.

Chemical equilibrium is defined as occurring when the quantities of reactants and products are at a 'steady state' and no longer shifting, but the reaction may still proceed forward and backward. The rate of forward reaction must equal the rate of backward reaction.

No matter is ever gained or lost during a chemical reaction; therefore the chemical equation must be balanced. This means that there must be the same number of atoms on both sides of the equation. Remember that the subscript numbers indicate the number of atoms in the elements. If there is no subscript, assume there is only one atom.

The number of molecules is shown by the number in front of an element or compound. If no number appears, assume that it is one molecule.

Many chemical reactions give off energy. Like matter, energy can change form but it can be neither created nor destroyed during a chemical reaction. This is the **law of conservation of energy.**

The **periodic table of elements** is an arrangement of the elements in rows and columns so that it is easy to locate elements with similar properties. The elements of the modern periodic table are arranged in numerical order by atomic number.

The **periods** are the rows down the left side of the table. They are called first period, second period, etc. The columns of the periodic table are called **groups**, or **families.** Elements in a family have similar properties.

There are three types of elements that are grouped by color: metals, nonmetals, and metalloids.

Element Key

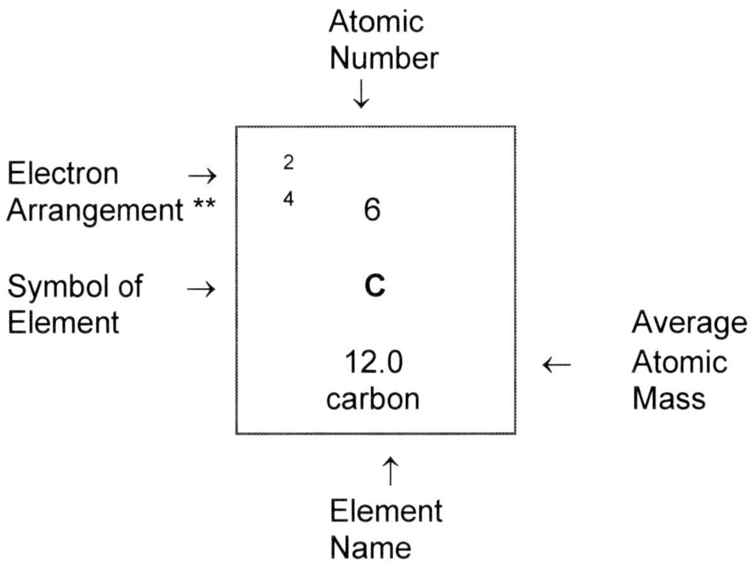

** Number of electrons on each level. Top number represents the innermost level.

An **acid** contains hydrogen ions (H+). Although it is never wise to taste a substance to identify it, acids have a sour taste. Vinegar and lemon juice are both acids, and acids occur in many foods in a weak state. Strong acids can burn skin and destroy materials. Common acids include:

Sulfuric acid (H_2SO_4)	-	Used in medicines, alcohol, dyes, and car batteries.
Nitric acid (HNO_3)	-	Used in fertilizers, explosives, cleaning materials.
Carbonic acid (H_2CO_3)	-	Used in soft drinks.
Acetic acid ($HC_2H_3O_2$)	-	Used in making plastics, rubber, photographic film, and as a solvent.

Bases have a bitter taste and the stronger ones feel slippery. Like acids, strong bases can be dangerous and should be handled carefully. All bases contain the hydroxyl ions (OH-). Many household cleaning products contain bases. Common bases include:

Sodium hydroxide	$NaOH$	-	Used in making soap, paper, vegetable oils, and refining petroleum.
Ammonium hydroxide	NH_4OH	-	Used in making deodorants, bleaching compounds, cleaning compounds.
Potassium hydroxide	KOH	-	Used in making soaps, drugs, dyes, alkaline batteries and purifying industrial gases.
Calcium	$Ca(OH)_2$	-	Used in making cement and plaster

COMPETENCY 12.0 UNDERSTAND PHYSICAL AND CHEMICAL CHANGES IN MATTER

Skill 12.1 Identify types and characteristics of physical (e.g., phase changes) and chemical (e.g., oxidation, combustion) changes in matter and factors that cause these changes

PHASE CHANGES:

Physical properties and chemical properties of matter describe the appearance or behavior of a substance. A **physical property** can be observed without changing the identity of a substance. For instance, you can describe the color, mass, shape, and volume of a book.

Matter constantly changes. A **physical change** is a change that does not produce a new substance. The freezing and melting of water is an example of physical change.

The phase of matter (solid, liquid, or gas) is identified by its shape and volume. A solid has a definite shape and volume. A liquid has a definite volume, but no shape. A gas has no shape or volume because it will spread out to occupy the entire space of whatever container it is in.

Energy is the ability to cause change in matter. Applying heat to a frozen liquid changes it from solid back to liquid. Continue heating it and it will boil and give off steam, a gas.

Evaporation is the change in phase from liquid to gas. Condensation is the change in phase from gas to liquid.

Oxidation reactions describe reactions where the molecule, atom, or ion loses an electron. **Reduction reactions** describe reactions where a molecule gains an electron.

Combustion (burning) is a sequence of chemical reactions between a fuel and an oxidizing agent (material that will gather an electron to itself, thereby stripping it from the original material) accompanied by the production of heat. Example:

$$2H_2 + O_2 \rightarrow 2H_2O + \text{heat}$$

Chemical properties describe the ability of a substance to be changed into new substances. Baking powder goes through a chemical change as it changes into carbon dioxide gas during the baking process.

A **chemical change** (or chemical reaction) is any change of a substance into one or more other substances. Burning materials turn into smoke; a seltzer tablet fizzes into gas bubbles.

Oxides are compounds that are formed when oxygen combines with another element. Rust is an oxide formed when oxygen combines with iron.

Skill 12.2 Apply the concept of conservation of matter

The principle of conservation states that certain measurable properties of an isolated system remain constant despite changes in the system. Two important principles of conservation are the conservation of mass and charge.

The principle of conservation of mass states that the total mass of a system is constant. Examples of conservation in mass in nature include the burning of wood, rusting of iron, and phase changes of matter. When wood burns, the total mass of the products, such as soot, ash, and gases, equals the mass of the wood and the oxygen that reacts with it. When iron reacts with oxygen, rust forms. The total mass of the iron-rust complex does not change. Finally, when matter changes phase, mass remains constant. Thus, when a glacier melts due to atmospheric warming, the mass of liquid water formed is equal to the mass of the glacier.

The principle of conservation of charge states that the total electrical charge of a closed system is constant. Thus, in chemical reactions and interactions of charged objects, the total charge does not change. Chemical reactions and the interaction of charged molecules are essential and common processes in living organisms and systems.

Skill 12.3 Identify properties of various states of matter and the energy transfer associated with changes in state

The **phase of matter** (solid, liquid, or gas) is identified by its shape and volume. A **solid** has a definite shape and volume. A **liquid** has a definite volume, but no shape. A **gas** has no shape or volume because it will spread out to occupy the entire space of whatever container it is in.

The kinetic theory states that matter consists of molecules, possessing kinetic energies, in continual random motion. The state of matter (solid, liquid, or gas) depends on the speed of the molecules and the amount of kinetic energy the molecules possess. The molecules of solid matter merely vibrate allowing strong intermolecular forces to hold the molecules in place. The molecules of liquid matter move freely and quickly throughout the body and the molecules of gaseous matter move randomly and at high speeds.

Matter changes state when energy is added or taken away. The addition of energy, usually in the form of heat, increases the speed and kinetic energy of the component molecules. Faster moving molecules more readily overcome the intermolecular attractions that maintain the form of solids and liquids. In conclusion, as the speed of molecules increases, matter changes state from solid to liquid to gas (melting and evaporation).

As matter loses heat energy to the environment, the speed of the component molecules decrease. Intermolecular forces have greater impact on slower moving molecules. Thus, as the speed of molecules decrease, matter changes from gas to liquid to solid (condensation and freezing).

Skill 12.4 Demonstrate an understanding of the predictable nature of chemical reactions

In one kind of chemical reaction, two elements combine to form a new substance. We can represent the reaction and the results in a chemical equation.

Carbon and oxygen form carbon dioxide. The equation can be written:

C	+	O_2	→	CO_2
1 atom of carbon	+	2 atoms of oxygen	→	1 molecule of carbon dioxide

In a second kind of chemical reaction, the molecules of a substance split forming two or more new substances. An electric current can split water molecules into hydrogen and oxygen gas.

$2H_2O$	→	$2H_2$	+	O_2
2 molecules of water	→	2 molecules of hydrogen	+	1 molecule of oxygen

The number of molecules is shown by the number in front of an element or compound. If no number appears, assume that it is 1 molecule.
A third kind of chemical reaction is when elements change places with each other. An example of one element taking the place of another is when iron changes places with copper in the compound copper sulfate:

$CuSo_4$	+	Fe	→	$FeSO_4$	+	Cu
copper sulfate	+	iron (steel wool)		iron sulfate		copper

Skill 12.5 Analyze how concentration, pressure, temperature, and catalysts affect chemical reactions (e.g., rate of reaction)

The rate of most simple reactions **increases with temperature** because a **greater fraction of molecules have the kinetic energy** required to overcome the reaction's activation energy. The chart below shows the effect of temperature on the distribution of kinetic energies in a sample of molecules. These curves are called **Maxwell-Boltzmann distributions**. The shaded areas represent the fraction of molecules containing sufficient kinetic energy for a reaction to occur. This area is larger at a higher temperature; so more molecules are above the activation energy and more molecules react per second.

This area is larger at a higher temperature so more molecules are above the activation energy and more molecules react per second.

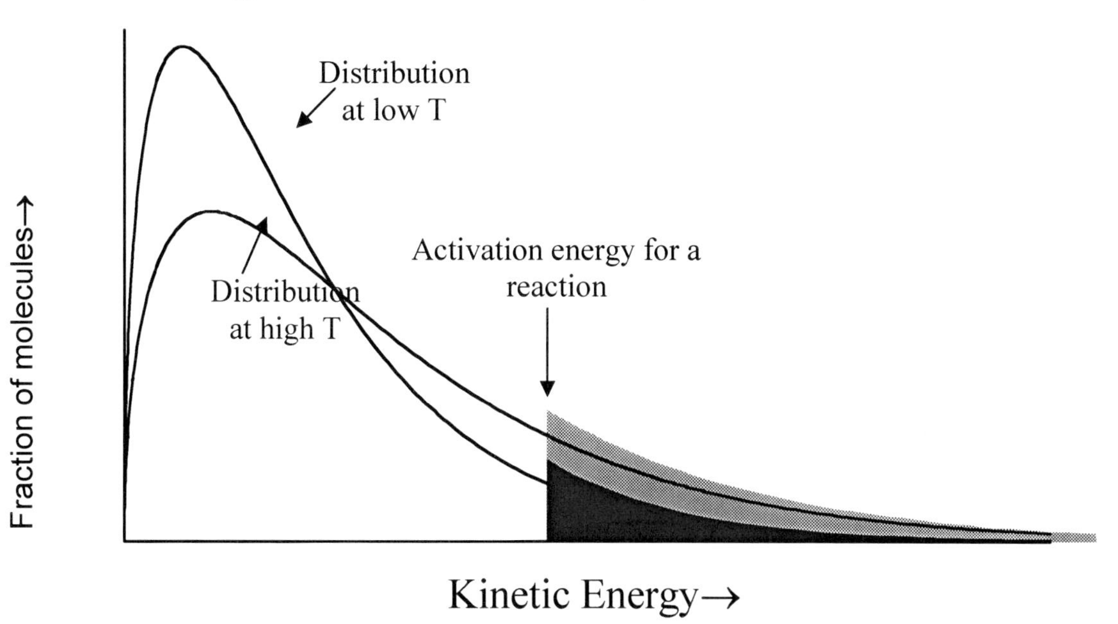

http://www.mhhe.com/physsci/chemistry/essentialchemistry/flash/activa2.swf provides an animated audio tutorial on energy diagrams.

Kinetic molecular theory may be applied to reaction rates in addition to physical constants like pressure. **Reaction rates increase with reactant concentration** because more reactant molecules are present and more are likely to collide with one another in a certain volume at higher concentrations. The nature of these relationships determines the rate law for the reaction. For ideal gases, the concentration of a reactant is its molar density, and this varies with pressure and temperature. Kinetic molecular theory also predicts that **reaction rate constants (values for *k*) increase with temperature** because of two reasons:

1. More reactant molecules will collide with each other per second.
2. These collisions will each occur at a higher energy that is more likely to overcome the activation energy of the reaction.

A **catalyst** is a material that increases the rate of a chemical reaction without changing itself permanently in the process. Catalysts provide an alternate reaction mechanism for the reaction to proceed in the forward and in the reverse direction. Therefore, **catalysts have no impact on the chemical equilibrium** of a reaction. They will not make a less favorable reaction more favorable.

Catalysts reduce the activation energy of a reaction. This is the amount of energy needed for the reaction to begin. Molecules with such low energies that they would have taken a long time to react will react more rapidly if a catalyst is present.

The impact of a catalyst may also be represented on an energy diagram. **A catalyst increases the rate of both the forward and reverse reactions by lowering the activation energy** for the reaction. Catalysts provide a different activated complex for the reaction at a lower energy state.

Biological catalysts are called **enzymes.**

Skill 12.6 Demonstrate an understanding of the use of symbolic equations to represent chemical changes and reactions

There are four kinds of chemical reactions:

In a **composition reaction**, two or more substances combine to form a compound.
$A + B \rightarrow AB$
i.e. silver and sulfur yield silver dioxide

In a **decomposition reaction**, a compound breaks down into two or more simpler substances.
$AB \rightarrow A + B$
i.e. water breaks down into hydrogen and oxygen

In a **single replacement reaction**, a free element replaces an element that is part of a compound.
$A + BX \rightarrow AX + B$
i.e. iron plus copper sulfate yields iron sulfate plus copper

In a **double replacement reaction**, parts of two compounds replace each other. In this case, the compounds seem to switch partners.
$AX + BY \rightarrow AY + BX$
i.e. sodium chloride plus mercury nitrate yields sodium nitrate plus mercury chloride

COMPETENCY 13.0 UNDERSTAND CONCEPTS RELATED TO ENERGY AND ENERGY TRANSFORMATION

Skill 13.1 Demonstrate knowledge of the forms of energy (e.g., heat, light, mechanical) and their characteristics

Heat is a measure of energy. If two objects that have different temperatures come into contact with each other, heat flows from the hotter object to the cooler.

Heat Capacity of an object is the amount of heat energy that it takes to raise the temperature of the object by one degree.

Heat capacity (C) per unit mass (m) is called **specific heat** (c):

$$c = \frac{C}{m} = \frac{Q / \Delta}{m}$$

Specific heats for many materials have been calculated and can be found in tables.

There are a number of ways that heat is measured. In each case, the measurement is dependent upon raising the temperature of a specific amount of water by a specific amount. These conversions of heat energy and work are called the **mechanical equivalent of heat**.

The **calorie** is the amount of energy that it takes to raise one gram of water one degree Celsius.

The **kilocalorie** is the amount of energy that it takes to raise one kilogram of water by one degree Celsius. Food calories are kilocalories.

In the International System of Units **(SI),** the calorie is equal to 4.184 **joules**.

A **British thermal unit (BTU)** = 252 calories = 1.054 kJ

Heat energy that is transferred into or out of a system is **heat transfer.** The temperature change is positive for a gain in heat energy and negative when heat is removed from the object or system.

The formula for heat transfer is $Q = mc\Delta T$ where Q is the amount of heat energy transferred, m is the amount of substance (in kilograms), c is the specific heat of the substance, and ΔT is the change in temperature of the substance. It is important to assume that the objects in thermal contact are isolated and insulated from their surroundings.

If a substance in a closed container loses heat, then another substance in the container must gain heat. A **calorimeter** uses the transfer of heat from one substance to another to determine the specific heat of the substance.

When an object undergoes a change of phase it goes from one physical state (solid, liquid, or gas) to another. For instance, water can go from liquid to solid (freezing) or from liquid to gas (boiling). The heat that is required to change from one state to the other is called **latent heat.**

The **heat of fusion** is the amount of heat that it takes to change from a solid to a liquid or the amount of heat released during the change from liquid to solid. The **heat of vaporization** is the amount of heat that it takes to change from a liquid to a gaseous state.

LIGHT:

Shadows illustrate one of the basic properties of **light**. Light travels in a straight line. If you put your hand between a light source and a wall, you will interrupt the light and produce a shadow.

When light hits a surface, it is **reflected.** The angle of the incoming light (angle of incidence) is the same as the angle of the reflected light (angle of reflection). It is this reflected light that allows you to see objects. You see the objects when the reflected light reaches your eyes.

Different surfaces reflect light differently. Rough surfaces scatter light in many different directions. A smooth surface reflects the light in one direction. If it is smooth and shiny (like a mirror) you see your image in the surface.

When light enters a different medium, it bends. This bending, or change of speed, is called **refraction**.

Light can be **diffracted**, or bent around the edges of an object. Diffraction occurs when light goes through a narrow slit. As light passes through it, the light bends slightly around the edges of the slit. You can demonstrate this by pressing your thumb and forefinger together, making a very thin slit between them. Hold them about 8 cm from your eye and look at a distant source of light. The pattern you observe is caused by the diffraction of light.

Light and other electromagnetic radiation can be polarized because the waves are transverse. The distinguishing characteristic of transverse waves is that they are perpendicular to the direction of the motion of the wave. Polarized light has vibrations confined to a single plane that is perpendicular to the direction of motion. Light is able to be polarized by passing it through special filters that block all vibrations except those in a single plane. By blocking out all but one place of vibration, polarized sunglasses cut down on glare.

Light can travel through thin fibers of glass or plastic without escaping the sides. Light on the inside of these fibers is reflected so that it stays inside the fiber until it reaches the other end. Such fiber optics are being used to carry telephone messages. Sound waves are converted to electric signals which are coded into a series of light pulses which move through the optical fiber until they reach the other end. At that time, they are converted back into sound.

The image that you see in a bathroom mirror is a virtual image because it only seems to be where it is. However, a curved mirror can produce a real image. A real image is produced when light passes through the point where the image appears. A real image can be projected onto a screen.

Cameras use a convex lens to produce an image on the film. A **convex lens** is thicker in the middle than at the edges. The image size depends upon the focal length (distance from the focus to the lens). The longer the focal length, the larger the image. A **converging lens** produces a real image whenever the object is far enough from the lens so that the rays of light from the object can hit the lens and be focused into a real image on the other side of the lens.

Eyeglasses can help correct deficiencies of sight by changing where the image seen is focused on the retina of the eye. If a person is nearsighted, the lens of his eye focuses images in front of the retina. In this case, the corrective lens placed in the eyeglasses will be concave so that the image will reach the retina. In the case of farsightedness, the lens of the eye focuses the image behind the retina. The correction will call for a convex lens to be fitted into the glass frames so that the image is brought forward into sharper focus.

MECHANICAL

Mechanical energy is the potential and kinetic energy of a mechanical system. Rolling balls, car engines, and body parts in motion exemplify mechanical energy.

Skill 13.2 Analyze the relationship between energy and matter and the ways in which they interact

Interacting objects in the universe constantly exchange and transform energy. Total energy remains the same, but the form of the energy readily changes. Energy often changes from kinetic (motion) to potential (stored) or potential to kinetic. In reality, available energy, energy that is easily utilized, is rarely conserved in energy transformations. Heat energy is an example of relatively "useless" energy often generated during energy transformations. Exothermic reactions release heat and endothermic reactions require heat energy to proceed. For example, the human body is notoriously inefficient in converting chemical energy from food into mechanical energy. The digestion of food is exothermic and produces substantial heat energy.

Skill 13.3 Demonstrate a basic understanding of the laws of thermodynamics (e.g., conservation of energy, entropy changes) and the processes of heat transfer (e.g., conduction, radiation, convection)

The relationship between heat, forms of energy, and work (mechanical, electrical, etc.) are the **Laws of Thermodynamics.** These laws deal strictly with systems in thermal equilibrium and not those within the process of rapid change or in a state of transition. Systems that are nearly always in a state of equilibrium are called **reversible systems.**

The First Law of Thermodynamics is a restatement of conservation of energy. The change in heat energy supplied to a system (Q) is equal to the sum of the change in the internal energy (U) and the change in the work done by the system against internal forces. $\Delta Q = \Delta U + \Delta W$

The Second Law of Thermodynamics is stated in two parts:

1. No machine is 100% efficient. It is impossible to construct a machine that only absorbs heat from a heat source and performs an equal amount of work because some heat will always be lost to the environment.

2. Heat can not spontaneously pass from a colder to a hotter object. An ice cube sitting on a hot sidewalk will melt into a little puddle, but it will never spontaneously cool and form the same ice cube. Certain events have a preferred direction called the **arrow of time.**

Entropy is the measure of how much energy or heat is available for work. Work occurs only when heat is transferred from hotter to cooler objects. Once this is done, no more work can be extracted. The energy is still being conserved, but is not available for work as long as the objects are the same temperature. Theory has it that, eventually, all things in the universe will reach the same temperature. If this happens, energy will no longer be usable.

Heat is transferred in three ways: **conduction, convection, and radiation.**

Conduction occurs when heat travels through a heated solid.

The transfer rate is the ratio of the amount of heat per amount of time it takes to transfer heat from area of an object to another. For example, if you place an iron pan on a flame, the handle will eventually become hot. How fast the handle gets too hot to handle is a function of the amount of heat and how long it is applied. Because the change in time is in the denominator of the function, the shorter the amount of time it takes to heat the handle, the greater the transfer rate.

Convection is heat transported by the movement of a heated substance. Warmed air rising from a heat source such as a fire or electric heater is a common example of convection. Convection ovens make use of circulating air to more efficiently cook food.

Radiation is heat transfer as the result of electromagnetic waves. The sun warms the earth by emitting radiant energy.

An example of all three methods of heat transfer occurs in the thermos bottle or Dewar flask. The bottle is constructed of double walls of Pyrex glass that have a space in between. Air is evacuated from the space between the walls and the inner wall is silvered. The lack of air between the walls lessens heat loss by convection and conduction. The heat inside is reflected by the silver, cutting down heat transfer by radiation. Hot liquids remain hotter and cold liquids remain colder for longer periods of time.

Skill 13.4 Demonstrate an understanding of energy transfer in physical systems (e.g., kinetic energy, potential energy, heat energy)

The Law of **Conservation of Energy** states that energy may neither be created nor destroyed. Therefore, the sum of all energies in the system is a constant.

Example:

The formula to calculate the potential energy is PE = mgh.
The mass of the ball = 20kg
The height, h = 0.4m
The acceleration due to gravity, g = 9.8 m/s^2

PE = mgh
PE = 20(.4)(9.8)
PE = 78.4J (Joules, units of energy)

The position of the ball on the left is where the Potential Energy (PE) = 78.4J resides while the Kinetic Energy (KE) = 0. As the ball is approaching the center position, the PE is decreasing while the KE is increasing. At exactly halfway between the left and center positions, the PE = KE.

The center position of the ball is where the Kinetic Energy is at its maximum while the Potential Energy (PE) = 0. At this point, theoretically, the entire PE has transformed into KE. Now the KE = 78.4J while the PE = 0.

The right position of the ball is where the Potential Energy (PE) is once again at its maximum and the Kinetic Energy (KE) = 0.
We can now say that:
PE + KE = 0
PE = -KE
The sum of PE and KE is the **total mechanical energy:**
Total Mechanical Energy = PE + KE

Skill 13.5 Apply the kinetic molecular model to explain the behavior of solids, liquids, and gases and to explain what happens during chemical reactions

Gas pressure results from molecular collisions with container walls. The number of molecules striking an area on the walls and the average kinetic energy per molecule are the only factors that contribute to pressure. A higher temperature increases speed and kinetic energy. There are more collisions at higher temperatures, but the average distance between molecules does not change, and thus density does not change in a sealed container.

Kinetic molecular theory explains how the pressure and temperature influences behavior of gases by making a few assumptions, namely:

1. The energies of intermolecular attractive and repulsive forces may be neglected.
2. The average kinetic energy of the molecules is proportional to absolute temperature.
3. Energy can be transferred between molecules during collisions and the collisions are elastic, so the average kinetic energy of the molecules doesn't change due to collisions.
4. The volume of all molecules in a gas is negligible compared to the total volume of the container.

Strictly speaking, molecules also contain some kinetic energy by rotating or experiencing other motions. The motion of a molecule from one place to another is called **translation**. Translational kinetic energy is the form that is transferred by collisions, and kinetic molecular theory ignores other forms of kinetic energy because they are not proportional to temperature.

COMPETENCY 14.0 UNDERSTAND ELECTRICITY, MAGNETS, AND ELECTROMAGNETISM AND THEIR ASSOCIATED FIELDS

Skill 14.1 Analyze the characteristics of static electricity and electric fields

Electrostatics is the study of stationary electric charges. A plastic rod that is rubbed with fur or a glass rod that is rubbed with silk will become electrically charged and will attract small pieces of paper. The charge on the plastic rod rubbed with fur is negative and the charge on glass rod rubbed with silk is positive.

Electrically charged objects share these characteristics:

1. Like charges repel one another.
2. Opposite charges attract each other.
3. Charge is conserved. A neutral object has no net change. If the plastic rod and fur are initially neutral, when the rod becomes charged by the fur a negative charge is transferred from the fur to the rod. The net negative charge on the rod is equal to the net positive charge on the fur.

Materials through which electric charges can easily flow are called **conductors**. Metals which are good conductors include silicon and boron. On the other hand, an **insulator** is a material through which electric charges do not move easily, if at all. Examples of insulators would be the nonmetal elements of the periodic table. A simple device used to indicate the existence of a positive or negative charge is called an **electroscope**. An electroscope is made up of a conducting knob and attached to it are very lightweight conducting leaves usually made of foil (gold or aluminum). When a charged object touches the knob, the leaves push away from each other because like charges repel. It is not possible to tell whether if the charge is positive or negative.

Charging by induction:
Touch a knob with a finger while a charged rod is nearby. The electrons will be repulsed and flow out of the electroscope through the hand. If the hand is removed while the charged rod remains close, the electroscope will retain the charge.

When an object is rubbed with a charged rod, the object will take on the same charge as the rod. However, charging by induction gives the object the opposite charge as that of the charged rod.

Grounding charge:
Charge can be removed from an object by connecting it to the earth through a conductor. The removal of static electricity by conduction is called **grounding**.

Skill 14.2 Demonstrate an understanding of the basic properties of electricity (e.g., current, voltage, resistance)

Current is the number of electrons per second that flow past a point in a circuit. Current is measured with a device called an ammeter. To use an ammeter, put it in series with the load you are measuring.

As electrons flow through a wire, they lose potential energy. Some is changed into heat energy because of resistance. **Resistance** is the ability of the material to oppose the flow of electrons through it. All substances have some resistance, even if they are a good conductor such as copper. This resistance is measured in units called **ohms**. A thin wire will have more resistance than a thick one because it will have less room for electrons to travel. In a thicker wire, there will be more possible paths for the electrons to flow. Resistance also depends upon the length of the wire. The longer the wire, the more resistance it will have. Potential difference, resistance, and current form a relationship know as **Ohm's Law**. Current **(I)** is measured in amperes and is equal to potential difference **(V)** divided by resistance **(R)**.

$$I = V / R$$

If you have a wire with resistance of 5 ohms and a potential difference of 75 volts, you can calculate the current by

$$I = 75 \text{ volts} / 5 \text{ ohms}$$
$$I = 15 \text{ amperes}$$

A current of 10 or more amperes will cause a wire to get hot. 22 amperes is about the maximum for a house circuit. Anything above 25 amperes can start a fire.

Skill 14.3 Interpret diagrams of simple electrical circuits and their characteristics

An **electric circuit** is a path along which electrons flow. A simple circuit can be created with a dry cell, wire, a bell, or a light bulb. When all are connected, the electrons flow from the negative terminal, through the wire to the device and back to the positive terminal of the dry cell. If there are no breaks in the circuit, the device will work. The circuit is closed. Any break in the flow will create an open circuit and cause the device to shut off.

The device (bell, bulb) is an example of a **load**. A load is a device that uses energy. Suppose that you also add a buzzer so that the bell rings when you press the buzzer button. The buzzer is acting as a **switch**. A switch is a device that opens or closes a circuit. Pressing the buzzer makes the connection complete and the bell rings. When the buzzer is not engaged, the circuit is open and the bell is silent.

A **series circuit** is one where the electrons have only one path along which they can move. When one load in a series circuit goes out, the circuit is open. An example of this is a set of Christmas tree lights that is missing a bulb. None of the bulbs will work.

A **parallel circuit** is one where the electrons have more than one path to move along. If a load goes out in a parallel circuit, the other load will still work because the electrons can still find a way to continue moving along the path.

Skill 14.4 Demonstrate knowledge of the characteristics of magnets and magnetic fields

Magnets have a north pole and a south pole. Like poles repel and opposing poles attract. A **magnetic field** is the space around a magnet where its force will affect objects. The closer you are to a magnet, the stronger the force. As you move away, the force becomes weaker.

Some materials act as magnets and some do not. This is because magnetism is a result of electrons in motion. The most important motion in this case is the spinning of the individual electrons. Electrons spin in pairs in opposite directions in most atoms. Each spinning electron has the magnetic field that it creates canceled out by the electron that is spinning in the opposite direction.

In an atom of iron, there are four unpaired electrons. The magnetic fields of these are not canceled out. Their fields add up to make a tiny magnet. There fields exert forces on each other setting up small areas in the iron called **magnetic domains** where atomic magnetic fields line up in the same direction.

You can make a magnet out of an iron nail by stroking the nail in the same direction repeatedly with a magnet. This causes poles in the atomic magnets in the nail to be attracted to the magnet. The tiny magnetic fields in the nail line up in the direction of the magnet. The magnet causes the domains pointing in its direction to grow in the nail. Eventually, one large domain results and the nail becomes a magnet.

A bar magnet has a north pole and a south pole. If you break the magnet in half, each piece will have a north and south pole.

The earth has a magnetic field. In a compass, a tiny, lightweight magnet is suspended and will line its south pole up with the North Pole magnet of the earth.

Skill 14.5 Demonstrate an understanding of the principles of electromagnetism

A magnet can be made out of a coil of wire by connecting the ends of the coil to a battery. When the current goes through the wire, the wire acts in the same way that a magnet does, it is called an **electromagnet**. The poles of the electromagnet will depend upon which way the electric current runs. An electromagnet can be made more powerful in three ways:

1. Make more coils.
2. Put an iron core (nail) inside the coils.
3. Use more battery power.

Telegraphs use electromagnets to work. When a telegraph key is pushed, current flows through a circuit, turning on an electromagnet which attracts an iron bar. The iron bar hits a sounding board which responds with a click. Release the key and the electromagnet turns off. Messages can be sent around the world in this way.

Scrap metal can be removed from waste materials by the use of a large electromagnet that is suspended from a crane. When the electromagnet is turned on, the metal in the pile of waste will be attracted to it. All other materials will stay on the ground.

Skill 14.6 Use the principles of electromagnetism to explain the operation of electric motors, generators, and transformers

An electric meter, such as the one found on the side of a house, contains an aluminum disk that sits directly in a magnetic field created by electricity flowing through a conductor. The more the electricity flows (current), the stronger the magnetic field is. The stronger the magnetic field, the faster the disk turns. The disk is connected to a series of gears that turn a dial. Meter readers record the number from that dial.

In a motor, electricity is used to create magnetic fields that oppose each other and cause the rotor to move. The wiring loops attached to the rotating shaft have a magnetic field opposing the magnetic field caused by the wiring in the housing of the motor that cannot move. The repelling action of the opposing magnetic fields turns the rotor.

A generator is a device that turns rotary mechanical energy into electrical energy. The process is based on the relationship between magnetism and electricity. As a wire or any other conductor moves across a magnetic field, an electric current occurs in the wire. The large generators used by the electric companies have a stationary conductor. A magnet attached to the end of a rotating shaft is positioned inside a stationary conducting ring that is wrapped with a long, continuous piece of wire. When the magnet rotates, it induces a small electric current in each section of wire as it passes. Each section of wire is a small, separate electric conductor. All the small currents of these individual sections add up to one large current, which is what is used for electric power.

A transformer is an electrical device that changes electricity of one voltage into another voltage, usually from high to low voltage. You can see transformers at the top of utility poles. A transformer uses two properties of electricity: first, magnetism surrounds an electric circuit and second, voltage is made when a magnetic field moves or changes strength. Voltage is a measure of the strength or amount of electrons flowing through a wire. If another wire is close to an electric current changing strength, the electric current will also flow into that other wire as the magnetism changes. A transformer takes in electricity at a higher voltage and lets it run through many coils wound around an iron core. The magnetism in the core alternates because the current is alternating. An output wire with fewer coils is also around the core. The changing magnetism makes a current in the output wire. Having fewer coils means less voltage, so the voltage is reduced.

COMPETENCY 15.0 UNDERSTAND FORCES AND MOTION

Skill 15.1 Compare and contrast types and characteristics of forces (e.g., gravitational, frictional) and how they affect the physical world

Dynamics is the study of the relationship between motion and the forces affecting motion. **Force** causes motion.

Mass and weight are not the same quantities. An object's **mass** gives it a reluctance to change its current state of motion. It is also the measure of an object's resistance to acceleration. The force that the earth's gravity exerts on an object with a specific mass is called the object's weight on earth. Weight is a force that is measured in Newtons. Weight (W) = mass times acceleration due to gravity (**W = mg**). To illustrate the difference between mass and weight, picture two rocks of equal mass on a balance scale. If the scale is balanced in one place, it will be balanced everywhere, regardless of the gravitational field. However, the weight of the stones would vary on a spring scale, depending upon the gravitational field. In other words, the stones would be balanced both on earth and on the moon. However, the weight of the stones would be greater on earth than on the moon.

Surfaces that touch each other have a certain resistance to motion. This resistance is **friction.**

1. The materials that make up the surfaces will determine the magnitude of the frictional force.
2. The frictional force is independent of the area of contact between the two surfaces.
3. The direction of the frictional force is opposite to the direction of motion.
4. The frictional force is proportional to the normal force between the two surfaces in contact.

Static friction describes the force of friction of two surfaces that are in contact but do not have any motion relative to each other, such as a block sitting on an inclined plane. **Kinetic friction** describes the force of friction of two surfaces in contact with each other when there is relative motion between the surfaces.

When an object moves in a circular path, a force must be directed toward the center of the circle in order to keep the motion going. This constraining force is called **centripetal force**. Gravity is the centripetal force that keeps a satellite circling the earth.

Push and pull –Pushing a volleyball or pulling a bowstring applies muscular force when the muscles expand and contract. Elastic force is when any object returns to its original shape (for example, when a bowstring is released).

Rubbing – Friction opposes the motion of one surface past another. Friction is common when slowing down a car or sledding down a hill.

Pull of gravity - is a force of attraction between two objects. Gravity questions can be raised not only on earth but also between planets and even black hole discussions.

Forces on objects at rest – The formula F= m/a is shorthand for force equals mass over acceleration. An object will not move unless the force is strong enough to move the mass. Also, there can be opposing forces holding the object in place. For instance, a boat may want to be forced by the currents to drift away but an equal and opposite force is a rope holding it to a dock.

Forces on a moving object - Overcoming inertia is the tendency of any object to oppose a change in motion. An object at rest tends to stay at rest. An object that is moving tends to keep moving.

Inertia and circular motion – The centripetal force is provided by the high banking of the curved road and by friction between the wheels and the road. This inward force that keeps an object moving in a circle is called centripetal force.

Skill 15.2 Analyze the motion of objects and the effects of forces on objects (e.g., resolving vectors, analyzing free-body diagrams)

The science of describing the motion of bodies is known as kinematics. The motion of bodies is described using words, diagrams, numbers, graphs, and equations.

The following words are used to describe motion: vectors, scalars, distance, displacement, speed, velocity, and acceleration.

The two categories of mathematical quantities that are used to describe the motion of objects are scalars and vectors. Scalars are quantities that are fully described by magnitude alone. Examples of scalars are 5m and 20 degrees Celsius. Vectors are quantities that are fully described by magnitude and direction. Examples of vectors are 30m/sec, and 5 miles north.

Distance is a scalar quantity that refers to how much ground an object has covered while moving. Displacement is a vector quantity that refers to the object's change in position.

Jamie walked 2 miles north, 4 miles east, 2 miles south, and then 4 miles west. In terms of distance, she walked 12 miles. However, there is no displacement because the directions cancelled each other out, and she returned to her starting position.

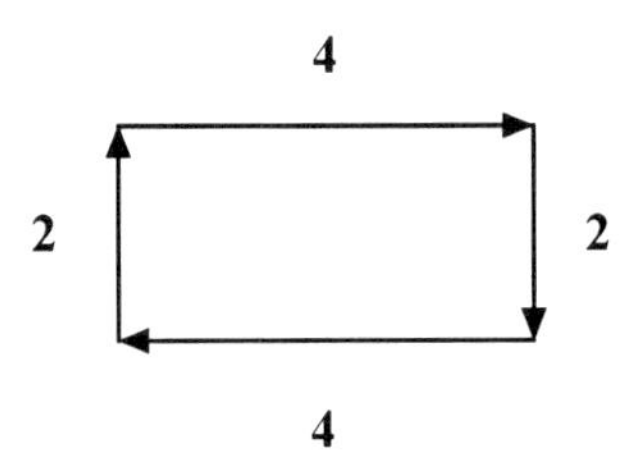

A free-body diagram is a diagram that shows the direction and relative magnitude of forces acting upon an object in a given situation. It is a special example of a vector diagram. The direction of the arrow indicates the direction in which the force is acting, and the size of the arrow indicates the magnitude of the force. Each arrow is labeled to represent the exact type of force. A box is used to represent the object and the force arrows are drawn outward from the center of the box in the directions they are acting.

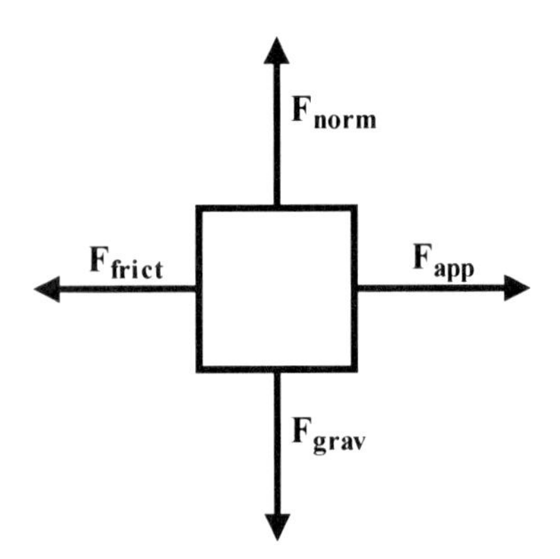

In this example, the object has four forces acting upon it, which is not always the case. Sometimes, there could be one, two, or three forces. The only rule is that all forces acting upon the object in that given situation should be depicted. Therefore, it is important that you be familiar with the various types of forces and be able to identify which forces are present in the situation.

Skill 15.3 Demonstrate an understanding of Newton's laws of motion and apply Newton's laws to a variety of practical problems (e.g., assessing frictional forces, determining forces acting on a pendulum)

NEWTONS LAWS:

Newton's laws of motion:

Newton's first law of motion is also called the law of inertia. It states that an object at rest will remain at rest and an object in motion will remain in motion at a constant velocity unless acted upon by an external force.

Newton's second law of motion states that if a net force acts on an object, it will cause the acceleration of the object. The relationship between force and motion is Force equals mass times acceleration. **($F = ma$).**

Newton's third law states that for every action there is an equal and opposite reaction. Therefore, if an object exerts a force on another object, that second object exerts an equal and opposite force on the first.

FRICTIONAL FORCES:

Surfaces that touch each other have a certain resistance to motion. This resistance is **friction.**

1. The materials that make up the surfaces will determine the magnitude of the frictional force.
2. The frictional force is independent of the area of contact between the two surfaces.
3. The direction of the frictional force is opposite to the direction of motion.
4. The frictional force is proportional to the normal force between the two surfaces in contact.

Static friction describes the force of friction of two surfaces that are in contact but do not have any motion relative to each other, such as a block sitting on an inclined plane. **Kinetic friction** describes the force of friction of two surfaces in contact with each other when there is relative motion between the surfaces.

FORCES ACTING ON A PENDULUM:

The motion of a pendulum is an example of mechanical energy conservation. A pendulum is made up of a bob attached to a pivot point by a string. The pendulum sweeps out a circular arc, moving back and forth in a periodic fashion. There are only two forces acting upon the bob. Gravity is one force, acting in a downward direction and doing work on the bob. However, gravity is a conservative, or internal force and does not change the total amount of mechanical energy of the bob. Tension is the other force acting upon the bob. It is an external force, which does not do work because it always acts in a direction perpendicular to the bob. The angle between the force of tension and its instantaneous displacement is always 90 degrees at all points in the trajectory of the bob. [F*d*cos(90°) = 0 Joules] Thus, the force of tension does not do work upon the bob.

There are no external forces doing work, so the total mechanical energy of the pendulum bob is conserved. The falling motion of the bob is accompanied by an increase in speed. As the bob loses height and potential energy, it gains speed and kinetic energy. The sum of potential energy and kinetic energy remains constant; the total of the two forms of mechanical energy is conserved.

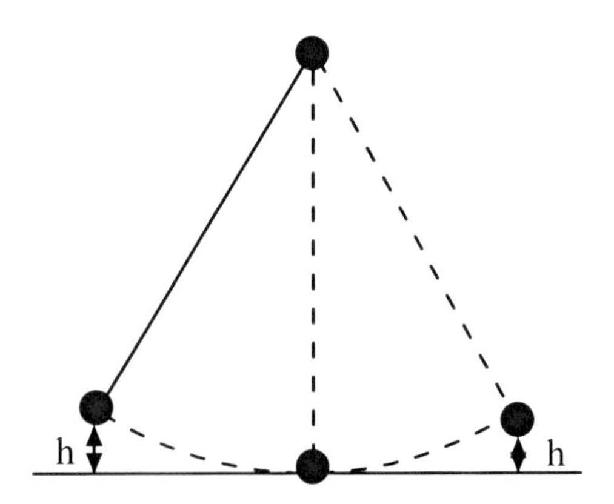

Skill 15.4 Demonstrate an understanding of the relationships of mass, force, motion, time, and position (e.g., inertia, momentum, velocity, acceleration)

Dynamics is the study of the relationship between motion and the forces affecting motion. Force causes motion.

Mass and weight are not the same quantities. An object's mass gives it a reluctance to change its current state of motion. It is also the measure of an object's resistance to acceleration. The force that the earth's gravity exerts on an object with a specific mass is called the object's weight on earth. Weight is a force that is measured in Newtons. Weight (W) = mass times acceleration due to gravity (W = mg). To illustrate the difference between mass and weight, picture two rocks of equal mass on a balance scale. If the scale is balanced in one place, it will be balanced everywhere, regardless of the gravitational field. However, the weight of the stones would vary on a spring scale, depending upon the gravitational field. In other words, the stones would be balanced both on earth and on the moon. However, the weight of the stones would be greater on earth than on the moon.

Inertia is the continuation of an object at rest to remain at rest. Conversely, momentum is the likelihood that an object in motion will remain in motion.

Speed is a scalar quantity that refers to how fast an object is moving (ex. the car was traveling 60 mi/hr). Velocity is a vector quantity that refers to the rate at which an object changes its position. In other words, velocity is speed with direction (ex. the car was traveling 60 mi/hr east).

Instantaneous Speed - speed at any given instant in time.

Average Speed - average of all instantaneous speeds, found simply by a distance/time ratio.

Acceleration is a vector quantity defined as the rate at which an object changes its velocity, where f represents the final velocity and i represents the initial velocity.

Since acceleration is a vector quantity, it always has a direction associated with it. The direction of the acceleration vector depends on whether the object is speeding up or slowing down and whether the object is moving in the positive or negative direction.

Skill 15.5 Compare and contrast characteristics of force, work, and power, and types and characteristics of simple machines

Mass and weight are not the same quantities. An object's **mass** gives it a reluctance to change its current state of motion. It is also the measure of an object's resistance to acceleration. The force that the earth's gravity exerts on an object with a specific mass is called the object's weight on earth. Weight is a force that is measured in Newtons. Weight (W) = mass times acceleration due to gravity (**W = mg**).

Simple Machines:

1. Inclined plane
2. Lever
3. Wheel and axle
4. Pulley

Work and energy:

Work is done on an object when an applied force moves through a distance.

Power is the work done divided by the amount of time that it took to do it. (Power = Work / time)

COMPETENCY 16.0 UNDERSTAND CHARACTERISTICS AND BEHAVIOR OF WAVES, SOUND, AND LIGHT

Skill 16.1 Identify types and characteristics of waves (e.g., amplitude, frequency, wavelength, speed) and their effects on properties of sound and light (e.g., pitch, color, spectrum)

The **pitch** of a sound depends on the **frequency** that the ear receives. High-pitched sound waves have high frequencies. High notes are produced by an object that is vibrating at a greater number of times per second than one that produces a low note.

The **intensity** of a sound is the amount of energy that crosses a unit of area in a given unit of time. The loudness of the sound is subjective and depends upon the effect on the human ear. Two tones of the same intensity but different pitches may appear to have different loudness. The intensity level of sound is measured in decibels. Normal conversation is about 60 decibels. A power saw is about 110 decibels.

The **amplitude** of a sound wave determines its loudness. Loud sound waves have large amplitudes. The larger the sound wave, the more energy is needed to create the wave.

Wavelength

Light, microwaves, x-rays, and TV and radio transmissions are all kinds of electromagnetic waves. They are all a wavy disturbance that repeats itself over a distance called the wavelength. Electromagnetic waves come in varying sizes and properties, by which they are organized in the electromagnetic spectrum. The electromagnetic spectrum is measured in frequency (f) in hertz and wavelength (λ) in meters. The frequency times the wavelength of every electromagnetic wave equals the speed of light (3.0 x 109 meters/second).

Roughly, the range of wavelengths of the electromagnetic spectrum are:

	λ		f	
Radio waves	10^{5} - 10^{-1}	meters	10^{3} - 10^{9}	hertz
Microwaves	10^{-1} - 10^{-3}	meters	10^{9} - 10^{11}	hertz
Infrared radiation	10^{-3} - 10^{-6}	meters	$10^{11.2}$ - $10^{14.3}$	hertz
Visible light	$10^{-6.2}$ $10^{-6.9}$	meters	$10^{14.3}$ - 10^{15}	hertz
Ultraviolet radiation	10^{-7} - 10^{-9}	meters	10^{15} - $10^{17.2}$	hertz
X-Rays	10^{-9} - 10^{-11}	meters	$10^{17.2}$ - 10^{19}	hertz
Gamma Rays	10^{-11} - 10^{-15}	meters	10^{19} - $10^{23.25}$	hertz

Radio waves are used for transmitting data. Common examples are television, cell phones, and wireless computer networks. Microwaves are used to heat food and deliver Wi-Fi service. Infrared waves are utilized in night vision goggles.

We are all familiar with visible light as the human eye is most sensitive to this wavelength range. UV light causes sunburns and would be even more harmful if most of it were not captured in the Earth's ozone layer. X-rays aid us in the medical field and gamma rays are most useful in the field of astronomy.

Skill 16.2 Compare and contrast the behavior of sound and light waves in various media (e.g., speed, transmission)

Sound waves need a medium in which to spread; light waves do not. The speed of any wave depends upon the elastic and inertial properties of the medium through which it travels. The density of a medium is an example of an **inertial property**. Sound usually travels faster in denser material. A sound wave will travel nearly three times as fast through helium than through air. On the other hand, the speed of light is slower in denser materials. The speed of light is slower in glass than in air. (The standard for the speed of light, c, is actually the speed of light in a vacuum, such as empty space.)

Elastic properties are properties related to the tendency of a medium to maintain its shape when acted upon by force or stress. Sound waves travel faster in solids than they do in liquids, and faster in liquids than they do in gases. The inertial factor would seem to indicate otherwise. However, the elastic factor has a greater influence on the speed of the wave.

When a wave strikes an object, some of the wave energy is reflected off the object, some of the energy goes into and is absorbed by the object, and some of the energy goes through the object. For example, sound waves can penetrate walls. However, sound waves from the air cannot penetrate water, and sound waves from water cannot penetrate the air. Light passes through some materials such as glass but not many other materials.

Skill 16.3 Analyze the behavior of light under various conditions (e.g., refraction, reflection, absorption, dispersion)

The place where one medium ends and another begins is called a **boundary**, and the manner in which a wave behaves when it reaches that boundary is called **boundary behavior**. The following principles apply to boundary behavior in waves:

1. wave speed is always greater in the less dense medium
2. wavelength is always greater in the less dense medium
3. wave frequency is not changed by crossing a boundary
4. the reflected pulse becomes inverted when a wave in a less dense medium is heading towards a boundary with a more dense medium
5. the amplitude of the incident pulse is always greater than the amplitude of the reflected pulse.

Reflection occurs when waves bounce off a barrier. The **law of reflection** states that when a ray of light reflects off a surface, the angle of incidence is equal to the angle of reflection.

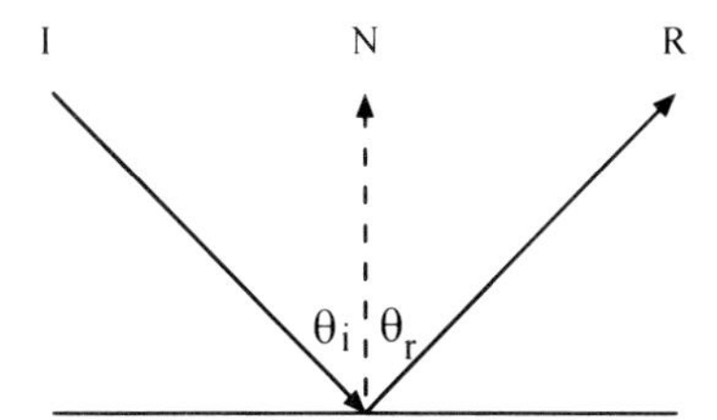

Line I represents the **incident ray**, the ray of light striking the surface. Line R is the **reflected ray**, the ray of light reflected off the surface. Line N is known as the **normal line**. It is a perpendicular line at the point of incidence that divides the angle between the incident ray and the reflected ray into two equal rays. The angle between the incident ray and the normal line is called the **angle of incidence**; the angle between the reflected ray and the normal line is called the **angle of reflection**.

Waves passing from one medium into another will undergo **refraction**, or bending. Accompanying this bending are a change in both speed and the wavelength of the waves.

In this example, light waves traveling through the air will pass through glass.

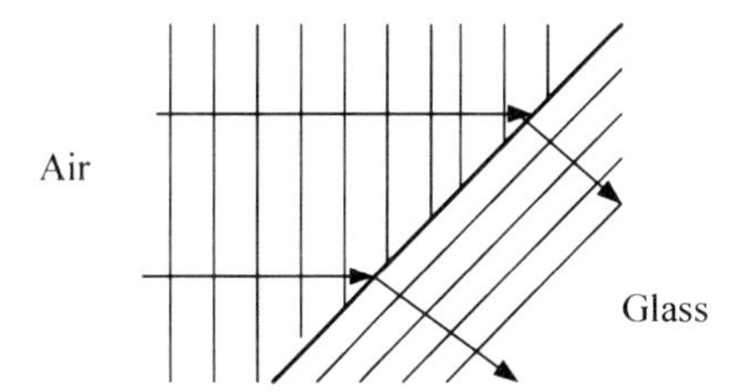

Refraction occurs only at the boundary. Once the wavefront passes across the boundary, it travels in a straight line.

Diffraction involves a change in direction of waves as they pass through an opening or around an obstacle in their path.

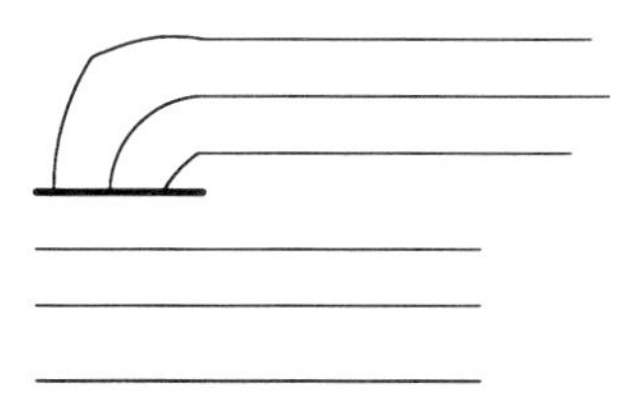

The amount of diffraction depends upon the wavelength. The amount of diffraction increases with increasing wavelength and decreases with decreasing wavelength. Sound and water waves exhibit this ability.

Skill 16.4 Demonstrate knowledge of phenomena related to sound and light (e.g., echoes, shadows, Doppler effect)

Shadows illustrate one of the basic properties of light. Light travels in a straight line. If you put your hand between a light source and a wall, you will interrupt the light and produce a shadow.

Light and other electromagnetic radiation can be polarized because the waves are transverse. The distinguishing characteristic of transverse waves is that they are perpendicular to the direction of the motion of the wave. Polarized light has vibrations confined to a single plane that is perpendicular to the direction of motion. Light is able to be polarized by passing it through special filters that block all vibrations except those in a single plane. By blocking out all but one place of vibration, polarized sunglasses cut down on glare.

Light can travel through thin fibers of glass or plastic without escaping the sides. Light on the inside of these fibers is reflected so that it stays inside the fiber until it reaches the other end. Such fiber optics are being used to carry telephone messages. Sound waves are converted to electric signals which are coded into a series of light pulses which move through the optical fiber until they reach the other end. At that time, they are converted back into sound.

Change in experienced frequency due to relative motion of the source of the sound is called the **Doppler Effect.** When a siren approaches, the pitch is high. When it passes, the pitch drops. As a moving sound source approaches a listener, the sound waves are closer together, causing an increase in frequency in the sound that is heard. As the source passes the listener, the waves spread out and the sound experienced by the listener is lower.

An **echo** is a wave that has been reflected by a medium, and returns to your ear. The delay between its reflection and your perception of its return is equal to the distance divided by the speed of sound.

SUBAREA IV. EARTH AND SPACE SCIENCE

COMPETENCY 17.0 UNDERSTAND THE STRUCTURE AND COMPOSITION OF EARTH AND THE NATURAL PROCESSES THAT SHAPE IT

Skill 17.1 Demonstrate knowledge of the composition and structure of Earth's interior

The interior of the Earth is divided in to three chemically distinct layers. Starting from the middle and moving towards the surface, these are the core, the mantle, and the crust. Much of what we know about the inner structure of the Earth has been inferred from various data. Subsequently, there is still some uncertainty about the composition and conditions in the Earth's interior.

Core
The **outer core** of the Earth begins about 3000 km beneath the surface and is a liquid, though far more viscous than that of the mantle. Even deeper, approximately 5000 km beneath the surface, is the solid **inner core**. The inner core has a radius of about 1200 km. Temperatures in the core exceed 4000 degrees Centigrade. Scientists agree that the core is extremely dense. This conclusion is based on the fact that the Earth is known to have an average density of 5515 kg/m^3 even though the material close to the surface has an average density of only 3000 kg/m^3. Therefore a denser core must exist. Additionally, it is hypothesized that when the Earth was forming, the densest material sank to the middle of the planet. Thus, it is not surprising that the core is about 80% iron. In fact, there is some speculation that the entire inner core is a single iron crystal, while the outer core is a mix of liquid iron and nickel.

Mantle
The Earth's mantle begins about 35 km beneath the surface and stretches all the way to 3000 km beneath the surface, where the outer core begins. Since the mantle stretches so far into the Earth's center, its temperature varies widely; near the boundary with the crust it is approximately 1000 degrees Centigrade, while near the outer core it may reach nearly 4000 degrees Centigrade. Within the mantle there are silicate rocks, which are rich in iron and magnesium. The silicate rocks exists as solids, but the high heat means they are ductile enough to "flow" over long time scales. In general, the mantle is semi-solid/plastic and the viscosity varies as pressures and temperatures change at varying depths.

Crust
It is not clear how long the Earth has actually had a solid crust; most of the rocks are less than 100 million years, though some are 4.4 billion years old. The crust of the earth is the outermost layer and continues down for between 5 and 70 km beneath the surface. Thin areas generally exist under ocean basins (oceanic crust) and thicker crust underlies the continents (continental crust). Oceanic crust is composed largely of iron magnesium silicate rocks, while continental crust is less dense and consists mainly of sodium potassium aluminum silicate rocks. The crust is the least dense layer of the Earth and so is rich in those materials that "floated" during Earth's formation. Additionally, some heavier elements that bound to lighter materials are present in the crust.

Interactions between the Layers
It is not the case that these layers exist as separate entities, with little interaction between them. For instance, it is generally believed that swirling of the iron-rich liquid in the outer core results in the Earth's magnetic field, which is readily apparent on the surface. Heat also moves out from the core to the mantle and crust. The core still retains heat from the formation of the Earth and additional heat is generated by the decay of radioactive isotopes. While most of the heat in our atmosphere comes from sun, radiant heat from the core does warm oceans and other large bodies of water.

Skill 17.2 Identify the physical and chemical properties of the lithosphere (e.g., rocks, minerals, soils) and analyze characteristics and processes of the rock cycle

Rocks are made up of minerals consisting of chemical elements or chemical compounds.

A chemical element is the simplest kind of substance. A chemical element can not be broken down into simpler substances by ordinary means. Gold, silver, oxygen and carbon are examples of chemical elements.

A chemical compound consists of two or more elements combined in fixed proportions by weight. Salt, water, table salt and natural gas are examples of chemical compounds.

Elements in the earth's crust are combinations of 88 different elements. Three common groups of minerals in the earth's crust are silicates, carbonates and oxides.

The two most common abundant elements in the earth's crust are oxygen and silicon.

The eight major elements in the earth's crust:

ELEMENT	SYMBOL	APPROX. % BY WEIGHT
Oxygen	O_2	47
Silicon	Si	28
Aluminum	Al	8
Iron	Fe	5
Calcium	Ca	4
Sodium	Na	3
Potasium	K	3
Magnesium	Mg	2

Minerals must adhere to five criteria. They must be (1) nonliving, (2) formed in nature, (3) solid in form, (4) their atoms form a crystalline pattern, and (5) its chemical composition is fixed within narrow limits.

There are over 3000 minerals in the earth's crust. Minerals are classified by composition. The major groups of minerals are silicates, carbonates, oxides, sulfides, sulfates, and halides. The largest group of minerals is the silicates. Silicates are made of silicon, oxygen and one or more other elements.

Soils are composed of particles of sand, clay, various minerals, tiny living organisms and humus, plus the decayed remains of plants and animals.

Soils are divided into three classes according to their texture. These classes are sandy soils, clay soils, and loamy soils.

Sandy soils are gritty, and their particles do not bind together firmly. Sandy soils are porous - water passes through them rapidly. Therefore, sandy soils do not hold much water.

Clay soils are smooth and greasy, their particles bind together firmly. Clay soils are moist and usually do not allow water to pass through easily.

Loamy soils feel somewhat like velvet and their particles clump together. Loamy soils are made up of sand, clay, and silt. Loamy soils holds water but some water can pass through.

Soils are grouped into three major types based upon their composition: pedalfers, pedocals and laterites.

Pedalfers form in the humid, temperate climate of the eastern United States. Pedalfer soils contain large amounts of iron oxide and alumunum-rich clays, making the soil a brown to reddish brown color. This soil supports forest type vegetation.

Pedocals are found in the western United States where the climate is dry and temperate. These soils are rich in calcium carbonate. This type of soil supports grasslands and brush vegetation.

Laterites are found where the climate is wet and tropical. Large amounts of water flows through this soil. Laterites are red-orange soils, rich in iron and alumunum oxides. There is little humus and this soil is not very fertile.

Rock Cycle

There is a great deal of interaction between the mantle and the crust. The slow convection of rocks in the mantle is responsible for the shifting of tectonic plates on the crust. Matter can also move between the layers as occurs during the **rock cycle**. Within the rock cycle, igneous rocks are formed when magma escapes from the mantle as lava during volcanic eruption. Rocks may also be forced back into the mantle, where the high heat and pressure recreate them as metamorphic rocks.

Skill 17.3 Identify major features of Earth's surface (e.g., mountains, oceans, plains, deep sea trenches) and analyze processes that produce changes in these features (e.g., weathering, erosion, deposition, plate tectonics)

Mountains

Orogeny is the term given to natural mountain building.

A mountain is terrain that has been raised high above the surrounding landscape by volcanic action, or some form of tectonic plate collisions. The plate collisions could be intercontinental or ocean floor collisions with a continental crust (subduction). The physical composition of mountains would include igneous, metamorphic, or sedimentary rocks; some may have rock layers that are tilted or distorted by plate collision forces.

There are many different types of mountains. The physical attributes of a mountain range depends upon the angle at which plate movement thrust layers of rock to the surface. Many mountains (Adirondacks, Southern Rockies) were formed along high angle faults.

Folded mountains (Alps, Himalayas) are produced by the folding of rock layers during their formation. The Himalayas are the highest mountains in the world and contain Mount Everest which rises almost 9 km above sea level. The Himalayas were formed when India collided with Asia. The movement which created this collision is still in process at the rate of a few centimeters per year.

Fault-block mountains (Utah, Arizona, and New Mexico) are created when plate movement produces tension forces instead of compression forces. The area under tension produces normal faults and rock along these faults is displaced upward.

Dome mountains are formed as magma tries to push up through the crust but fails to break the surface. Dome mountains resemble a huge blister on the earth's surface.

Upwarped mountains (Black Hills of South Dakota) are created in association with a broad arching of the crust. They can also be formed by rock thrust upward along high angle faults.

Volcanism is the term given to the movement of magma through the crust and its emergence as lava onto the earth's surface. Volcanic mountains are built up by successive deposits of volcanic materials.

An active volcano is one that is presently erupting or building to an eruption. A dormant volcano is one that is between eruptions but still shows signs of internal activity that might lead to an eruption in the future. An extinct volcano is said to be no longer capable of erupting. Most of the world's active volcanoes are found along the rim of the Pacific Ocean, which is also a major earthquake zone. This curving belt of active faults and volcanoes is often called the Ring of Fire.

The world's best known volcanic mountains include: Mount Etna in Italy and Mount Kilimanjaro in Africa. The Hawaiian Islands are actually the tops of a chain of volcanic mountains that rise from the ocean floor.

There are three types of volcanic mountains: shield volcanoes, cinder cones and composite volcanoes.

Shield Volcanoes are associated with quiet eruptions. Lava emerges from the vent or opening in the crater and flows freely out over the earth's surface until it cools and hardens into a layer of igneous rock. A repeated lava flow builds this type of volcano into the largest volcanic mountain. Mauna Loa found in Hawaii, is the largest volcano on earth.

Cinder Cone Volcanoes are associated with explosive eruptions as lava is hurled high into the air in a spray of droplets of various sizes. These droplets cool and harden into cinders and particles of ash before falling to the ground. The ash and cinder pile up around the vent to form a steep, cone-shaped hill called the cinder cone. Cinder cone volcanoes are relatively small but may form quite rapidly.

Composite Volcanoes are described as being built by both lava flows and layers of ash and cinders. Mount Fuji in Japan, Mount St. Helens in Washington, USA and Mount Vesuvius in Italy are all famous composite volcanoes.

Mechanisms of producing mountains

Mountains are produced by different types of mountain-building processes. Most major mountain ranges are formed by the processes of folding and faulting.

Folded Mountains are produced by the folding of rock layers. Crustal movements may press horizontal layers of sedimentary rock together from the sides, squeezing them into wavelike folds. Up-folded sections of rock are called anticlines; down-folded sections of rock are called synclines. The Appalachian Mountains are an example of folded mountains with long ridges and valleys in a series of anticlines and synclines formed by folded rock layers.

Faults are fractures in the earth's crust which have been created by either tension or compression forces transmitted through the crust. These forces are produced by the movement of separate blocks of crust.

Faultings are categorized on the basis of the relative movement between the blocks on both sides of the fault plane. The movement can be horizontal, vertical or oblique.

A dip-slip fault occurs when the movement of the plates is vertical and opposite. The displacement is in the direction of the inclination, or dip, of the fault. Dip-slip faults are classified as normal faults when the rock above the fault plane moves down relative to the rock below.

Reverse faults are created when the rock above the fault plane moves up relative to the rock below. Reverse faults having a very low angle to the horizontal are also referred to as thrust faults.

Faults in which the dominant displacement is horizontal movement along the trend or strike (length) of the fault are called **strike-slip faults**. When a large strike-slip fault is associated with plate boundaries it is called a transform fault. The San Andreas Fault in California is a well-known transform fault.

Faults that have both vertical and horizontal movement are called **oblique-slip faults.**

When lava cools, igneous rock is formed. This formation can occur either above ground or below ground.

Intrusive rock includes any igneous rock that was formed below the earth's surface. Batholiths are the largest structures of intrusive type rock and are composed of near granite materials; they are the core of the Sierra Nevada Mountains.

Extrusive rock includes any igneous rock that was formed at the earth's surface.

Dikes are old lava tubes formed when magma entered a vertical fracture and hardened. Sometimes magma squeezes between two rock layers and hardens into a thin horizontal sheet called a sill. A laccolith is formed in much the same way as a sill, but the magma that creates a laccolith is very thick and does not flow easily. It pools and forces the overlying strata creating an obvious surface dome.

A **caldera** is normally formed by the collapse of the top of a volcano. This collapse can be caused by a massive explosion that destroys the cone and empties most if not all of the magma chamber below the volcano. The cone collapses into the empty magma chamber forming a caldera.

An inactive volcano may have magma solidified in its pipe. This structure, called a volcanic neck, is resistant to erosion and today may be the only visible evidence of the past presence of an active volcano.

When lava cools, igneous rock is formed. This formation can occur either above ground or below ground.

Glaciation

A continental glacier covered a large part of North America during the most recent ice age. Evidence of this glacial coverage remains as abrasive grooves, large boulders from northern environments dropped in southerly locations, glacial troughs created by the rounding out of steep valleys by glacial scouring, and the remains of glacial sources called cirques that were created by frost wedging the rock at the bottom of the glacier. Remains of plants and animals found in warm climate have been discovered in the moraines and out wash plains help to support the theory of periods of warmth during the past ice ages.

The Ice Age began about 2 -3 million years ago. This age saw the advancement and retreat of glacial ice over millions of years. Theories relating to the origin of glacial activity include Plate Tectonics, where it can be demonstrated that some continental masses, now in temperate climates, were at one time blanketed by ice and snow. Another theory involves changes in the earth's orbit around the sun, changes in the angle of the earth's axis, and the wobbling of the earth's axis. Support for the validity of this theory has come from deep ocean research that indicates a correlation between climatic sensitive microorganisms and the changes in the earth's orbital status.

About 12,000 years ago, a vast sheet of ice covered a large part of the northern United States. This huge, frozen mass had moved southward from the northern regions of Canada as several large bodies of slow-moving ice, or glaciers. A time period in which glaciers advance over a large portion of a continent is called an ice age. A glacier is a large mass of ice that moves or flows over the land in response to gravity. Glaciers form among high mountains and in other cold regions.

There are two main types of glaciers: valley glaciers and continental glaciers. Erosion by valley glaciers is characteristic of U-shaped erosion. They produce sharp peaked mountains such as the Matterhorn in Switzerland. Erosion by continental glaciers often rides over mountains in their paths leaving smoothed, rounded mountains and ridges.

Oceans

The ocean floor has many of the same features that are found on land. The ocean floor has higher mountains than present on land, extensive plains and deeper canyons than present on land. Oceanographers have named different parts of the ocean floor according to their structure. The major parts of the ocean floor are:

The **continental shelf** is the sloping part of the continent that is covered with water extending from the shoreline to the continental slope.

The **continental slope** is the steeply sloping area that connects the continental shelf and the deep-ocean floor.

The **continental rise** is the gently sloping surface at the base of the continental slope.

The **abyssal plains** are the flat, level parts of the ocean floor.

A **seamount** is an undersea volcano peak that is at least 1000 m above the ocean floor.

Guyot - A submerged flat-topped seamount

Mid ocean-ridges are continuous undersea mountain chains that are found mostly in the middle portions of the oceans.

Ocean trenches are long, elongated narrow troughs or depressions formed where ocean floors collide with another section of ocean floor or continent. The deepest trench in the Pacific Ocean is the Marianas Trench which is about 11 km deep.

Erosion

Erosion is the inclusion and transportation of surface materials by another moveable material, usually water, wind, or ice. The most important cause of erosion is running water. Streams, rivers, and tides are constantly at work removing weathered fragments of bedrock and carrying them away from their original location.

A stream erodes bedrock by the grinding action of the sand, pebbles and other rock fragments. This grinding against each other is called abrasion.

Streams also erode rocks by dissolving or absorbing their minerals. Limestone and marble are readily dissolved by streams.

The breaking down of rocks at or near to the earth's surface is known as **weathering**. Weathering breaks down these rocks into smaller and smaller pieces. There are two types of weathering: physical weathering and chemical weathering.

Physical weathering is the process by which rocks are broken down into smaller fragments without undergoing any change in chemical composition. Physical weathering is mainly caused by the freezing of water, the expansion of rock, and the activities of plants and animals.

Frost wedging is the cycle of daytime thawing and refreezing at night. This cycle causes large rock masses, especially the rocks exposed on mountain tops, to be broken into smaller pieces.

The peeling away of the outer layers from a rock is called exfoliation. Rounded mountain tops are called exfoliation domes and have been formed in this way.

Chemical weathering is the breaking down of rocks through changes in their chemical composition. An example would be the change of feldspar in granite to clay. Water, oxygen, and carbon dioxide are the main agents of chemical weathering. When water and carbon dioxide combine chemically, they produce a weak acid that breaks down rocks.

Deposition, also known as sedimentation, is the term for the process by which material from one area is slowly deposited into another area. This is usually due to the movement of wind, water, or ice containing particles of matter. When the rate of movement slows down, particles filter out and remain behind, causing a build up of matter. Note that this is a result of matter being eroded and removed from another site.

Plate Tectonics

Data obtained from many sources led scientists to develop the theory of plate tectonics. This theory is the most current model that explains not only the movement of the continents, but also the changes in the earth's crust caused by internal forces.

Plates are rigid blocks of earth's crust and upper mantle. These rigid solid blocks make up the lithosphere. The earth's lithosphere is broken into nine large sections and several small ones. These moving slabs are called plates. The major plates are named after the continents they are "transporting."

The plates float on and move with a layer of hot, plastic-like rock in the upper mantle. Geologists believe that the heat currents circulating within the mantle cause this plastic zone of rock to slowly flow, carrying along the overlying crustal plates.

Movement of these crustal plates creates areas where the plates diverge as well as areas where the plates converge. A major area of divergence is located in the Mid-Atlantic. Currents of hot mantle rock rise and separate at this point of divergence creating new oceanic crust at the rate of 2 to 10 centimeters per year. Convergence is when the oceanic crust collides with either another oceanic plate or a continental plate. The oceanic crust sinks forming an enormous trench and generating volcanic activity. Convergence also includes continent to continent plate collisions. When two plates slide past one another a transform fault is created.

These movements produce many major features of the earth's surface, such as mountain ranges, volcanoes, and earthquake zones. Most of these features are located at plate boundaries, where the plates interact by spreading apart, pressing together, or sliding past each other. These movements are very slow, averaging only a few centimeters a year.

Boundaries form between spreading plates where the crust is forced apart in a process called rifting. Rifting generally occurs at mid-ocean ridges. Rifting can also take place within a continent, splitting the continent into smaller landmasses that drift away from each other, thereby forming an ocean basin between them. The Red Sea is a product of rifting. As the seafloor spreading takes place, new material is added to the inner edges of the separating plates. In this way the plates grow larger, and the ocean basin widens. This is the process that broke up the super continent Pangaea and created the Atlantic Ocean.

Boundaries between plates that are colliding are zones of intense crustal activity. When a plate of ocean crust collides with a plate of continental crust, the more dense oceanic plate slides under the lighter continental plate and plunges into the mantle. This process is called subduction, and the site where it takes place is called a subduction zone. A subduction zone is usually seen on the sea-floor as a deep depression called a **trench**.

The crustal movement which is identified by plates sliding sideways past each other produces a plate boundary characterized by major faults that are capable of unleashing powerful earth-quakes. The San Andreas Fault forms such a boundary between the Pacific Plate and the North American Plate.

Skill 17.4 Demonstrate knowledge of how fossils form and how fossils provide evidence of complexity and diversity over time

A fossil is the remains or trace of an ancient organism that has been preserved naturally in the Earth's crust. Sedimentary rocks usually are rich sources of fossil remains. Those fossils found in layers of sediment were embedded in the slowly forming sedimentary rock strata. The oldest fossils known are the traces of 3.5 billion year old bacteria found in sedimentary rocks. Few fossils are found in metamorphic rock and virtually none found in igneous rocks. The magma is so hot that any organism trapped in the magma is destroyed.

The fossil remains of a woolly mammoth embedded in ice were found by a group of Russian explorers. However, the best-preserved animal remains have been discovered in natural tar pits. When an animal accidentally fell into the tar, it became trapped, sinking to the bottom. Preserved bones of the saber-toothed cat have been found in tar pits.

Prehistoric insects have been found trapped in ancient amber or fossil resin that was excreted by some extinct species of pine trees.

Fossil molds are the hollow spaces in a rock previously occupied by bones or shells. A fossil cast is a fossil mold that fills with sediments or minerals that later hardens forming a cast.

Fossil tracks are the imprints in hardened mud left behind by birds or animals.

Infer the history of an area using geologic evidence

The determination of the age of rocks by cataloging their composition has been outmoded since the middle 1800s. Today a sequential history can be determined by the fossil content (principle of fossil succession) of a rock system as well as its superposition within a range of systems. This classification process was termed stratigraphy and permitted the construction of a Geologic Column in which rock systems are arranged in their correct chronological order.

Skill 17.5 Demonstrate knowledge of methods and techniques for locating points and features on Earth's surface (e.g., rectilinear coordinate systems, topographic maps, global positioning systems)

A system of imaginary lines has been developed that helps people describe exact locations on Earth. Looking at a globe of Earth, you will see lines drawn on it. The equator is drawn around Earth halfway between the North and South Poles. Latitude is a term used to describe distance in degrees north or south of the equator. Lines of latitude are drawn east and west parallel to the equator. Degrees of latitude range from 0 at the equator to 90 at either the North Pole or South Pole. Lines of latitude are also called parallels.

Lines drawn north and south at right angles to the equator and from pole to pole are called meridians. Longitude is a term used to describe distances in degrees east or west of a 0° meridian. The prime meridian is the 0° meridian and it passes through Greenwich, England.

Time zones are determined by longitudinal lines. Each time zone represents one hour. Since there are 24 hours in one complete rotation of the Earth, there are 24 international time zones. Each time zone is roughly 15° wide. While time zones are based on meridians, they do not strictly follow lines of longitude. Time zone boundaries are subject to political decisions and have been moved around cities and other areas at the whim of the electorate.

The International Date Line is the 180° meridian and it is on the opposite side of the world from the prime meridian. The International Date Line is one-half of one day or 12 time zones from the prime meridian. If you were traveling west across the International Date Line, you would lose one day. If you were traveling east across the International Date Line, you would gain one day.

Topographic Maps: Principles of contouring

A contour line is a line on a map representing an imaginary line on the ground that has the same elevation above sea level along its entire length. Contour intervals usually are given in even numbers or as a multiple of five. In mapping mountains, a large contour interval is used. Small contour intervals may be used where there are small differences in elevation.

Relief describes how much variation in elevation an area has. Rugged or high relief, describes an area of many hills and valleys. Gentle or low relief describes a plain area or a coastal region. Five general rules should be remembered in studying contour lines on a map.

1. Contour lines close around hills and basins or depressions. Hachure lines are used to show depressions. Hachures are short lines placed at right angles to the contour line and they always point toward the lower elevation. A contour line that has hachures is called a depression contour.

2. Contours lines never cross. Contour lines are sometimes very close together. Each contour line represents a certain height above sea level.

3. Contour lines appear on both sides of an area where the slope reverses direction. Contour lines show where an imaginary horizontal plane would slice through a hillside or cut both sides of a valley.

4. Contours lines form V's that point upstream when they cross streams. Streams cut beneath the general elevation of the land surface, and contour lines follow a valley.

5. All contours lines either close (connect) or extend to the edge of the map. No map is large enough to have all its contour lines close.

Interpret maps and imagery

Like photographs, maps readily display information that would be impractical to express in words. Maps that show the shape of the land are called topographic maps. Topographic maps, which are also referred to as quadrangles, are generally classified according to publication scale. Relief refers to the difference in elevation between any two points. Maximum relief refers to the difference in elevation between the high and low points in the area being considered. Relief determines the contour interval, which is the difference in elevation between succeeding contour lines that are used on topographic maps.

Map scales express the relationship between distance or area on the map to the true distance or area on the earth's surface. It is expressed as so many feet (miles, meters, km, or degrees) per inch (cm) of map.

Global positioning systems (GPS):

One relatively recent technology that can be used to locate points on the Earth is the **global positioning system (GPS)**. Over 20 GPS satellites broadcast signals that allow GPS receivers to obtain exact longitude, latitude, and altitude data. In the past 2 decades, GPS technology has become invaluable for navigation, military use, surveyors, and outdoor enthusiasts. Though GPS will not replace maps, these tools are incredibly powerful when used together.

COMPETENCY 18.0 UNDERSTAND CHARACTERISTICS OF THE ATMOSPHERE, WEATHER, AND CLIMATE

Skill 18.1 Demonstrate knowledge of the basic composition, structure, and properties of the atmosphere

Dry air is composed of three basic components; dry gas, water vapor, and solid particles (dust from soil, etc.).

The most abundant dry gases in the atmosphere are:

(N_2)	Nitrogen	78.09 %	makes up about 4/5 of gases in atmosphere
(O_2)	Oxygen	20.95 %	
(Ar)	Argon	0.93 %	
(CO_2)	Carbon Dioxide	0.03 %	

The atmosphere is divided into four main layers based on temperature. These layers are labeled Troposphere, Stratosphere, Mesosphere, Thermosphere.

Troposphere - this layer is the closest to the earth's surface and all weather phenomena occurs here as it is the layer with the most water vapor and dust. Air temperature decreases with increasing altitude. The average thickness of the Troposphere is 7 miles (11 km).

Stratosphere - this layer contains very little water, clouds within this layer are extremely rare. The Ozone layer is located in the upper portions of the stratosphere. Air temperature is fairly constant but does increase somewhat with height due to the absorption of solar energy and ultra violet rays from the ozone layer.

Mesosphere - air temperature again decreases with height in this layer. It is the coldest layer with temperatures in the range of -100^0 C at the top..

Thermosphere - extends upward into space. Oxygen molecules in this layer absorb energy from the sun, causing temperatures to increase with height. The lower part of the thermosphere is called the Ionosphere. Here charged particles or ions and free electrons can be found. When gases in the Ionosphere are excited by solar radiation, the gases give off light and glow in the sky. These glowing lights are called the Aurora Borealis in the Northern Hemisphere and Aurora Australis in Southern Hemisphere. The upper portion of the Thermosphere is called the Exosphere. Gas molecules are very far apart in this layer. Layers of Exosphere are also known as the Van Allen Belts and are held together by earth's magnetic field.

Skill 18.2 Demonstrate an understanding of the processes of energy transfer in the atmosphere (e.g., convection, radiation, phase changes of water)

Energy is transferred in Earth's atmosphere in three ways. Earth gets most of its energy from the sun in the form of waves. This transfer of energy by waves is termed **radiation**. The transfer of thermal energy through matter by actual contact of molecules is called **conduction**. For example, heated rocks and sandy beaches transfer heat to the surrounding air. The transfer of thermal energy due to air density differences is called **convection**. Convection currents circulate in a constant exchange of cold dense air for less dense warm air.

Carbon dioxide in the atmosphere absorbs energy from the sun. Carbon dioxide also blocks the direct escape of energy from the Earth's surface. This process by which heat is trapped by gases, water vapor and other gases in the Earth's atmosphere is called the **Greenhouse Effect**.

Most of the Earth's water is found in the oceans and lakes. Through the **water cycle**, water evaporates into the atmosphere and condenses into clouds. Water then falls to the Earth in the form of precipitation, returning to the oceans and lakes on falling on land. Water on the land may return to the oceans and lakes as runoff or seep from the soil as groundwater.

Skill 18.3 Identify types and characteristics of clouds and the processes of cloud formation and precipitation

CLOUD TYPES:

Cirrus clouds - White and feathery; high in the sky

Cumulus – thick, white, fluffy

Stratus – layers of clouds cover most of the sky

Nimbus – heavy, dark clouds that represent thunderstorm clouds

Variation on the clouds mentioned above include **Cumulo-nimbus** and **Strato-nimbus.**

The air temperature at which water vapor begins to condense is called the **dew point.**

Relative humidity is the actual amount of water vapor in a certain volume of air compared to the maximum amount of water vapor this air could hold at a given temperature.

FORMATION:

Condensation or the removal of water above the Earth's surface results in the formation of clouds. Generally, clouds develop in any air mass that becomes saturated or has a relative humidity of 100%. Certain processes that cool the temperature of an air mass to its dew point or frost point can cause saturation. There are four processes or any combination of these processes that create saturation and cause clouds to form:

1. Orographic uplift occurs when elevated land forces air to rise.
2. Convectional lifting is the result of surface heating of air at ground level. If enough heating occurs, the air rises, expands, and cools.
3. Convergence or frontal lifting occurs when two air masses come together. One of the air masses is usually warm and moist, while the other is cool and dry.
4. Radiative cooling usually occurs at night when the sun is no longer heating the ground and the surrounding air. The ground and the air begin to cool forming fog.

Skill 18.4 Analyze characteristics of large-scale and local weather systems (e.g., air masses, fronts, upper-level wind patterns) and the causes and effects of severe weather events (e.g., tornadoes, thunderstorms, blizzards)

AIR MASSES:

Air masses moving toward or away from the Earth's surface are called air currents. Air moving parallel to Earth's surface is called **wind**. Weather conditions are generated by winds and air currents carrying large amounts of heat and moisture from one part of the atmosphere to another. Wind speeds are measured by instruments called anemometers.

FRONTS:

Fronts are the leading edges of air masses which have different density (air temperature and/or humidity) than the area it is passing into. When a front passes over an area, the area often experiences changes in temperature, moisture, wind speed/direction, atmospheric pressure, and even precipitation. Cold fronts are often associated with low pressure systems and bring cold, dry air. Warm fronts often bring warm, moist air which may feel tropical. Fronts are guided by winds and travel from west to east. Fronts may be manipulated by mountains and large bodies of water.

UPPER-LEVEL WIND PATTERNS:

The wind belts in each hemisphere consist of convection cells that encircle Earth like belts. There are three major wind belts on Earth: (1) trade winds (2) prevailing westerlies, and (3) polar easterlies. Wind belt formation depends on the differences in air pressures that develop in the doldrums, the horse latitudes, and the polar regions. The Doldrums surround the equator. Within this belt heated air usually rises straight up into Earth's atmosphere. The Horse latitudes are regions of high barometric pressure with calm and light winds and the Polar regions contain cold dense air that sinks to the Earth's surface.

TYPES OF STORMS:

A **thunderstorm** is a brief, local storm produced by the rapid upward movement of warm, moist air within a cumulo-nimbus cloud. Thunderstorms always produce lightning and thunder, and are accompanied by strong wind gusts and heavy rain or hail.

A severe storm with swirling winds that may reach speeds of hundreds of km per hour is called a **tornado**. Such a storm is also referred to as a "twister". The sky is covered by large cumulo-nimbus clouds and violent thunderstorms; a funnel-shaped swirling cloud may extend downward from a cumulo-nimbus cloud and reach the ground. Tornadoes are storms that leave a narrow path of destruction on the ground.

A swirling, funnel-shaped cloud that extends downward and touches a body of water is called a **waterspout.**

Hurricanes are storms that develop when warm, moist air carried by trade winds rotate around a low-pressure "eye". A large, rotating, low-pressure system accompanied by heavy precipitation and strong winds is called a tropical cyclone (better known as a hurricane). In the Pacific region, a hurricane is called a typhoon.

Storms that occur only in the winter are known as blizzards or ice storms. A **blizzard** is a storm with strong winds, blowing snow and frigid temperatures. An **ice storm** consists of falling rain that freezes when it strikes the ground, covering everything with a layer of ice.

Skill 18.5 Identify the characteristics and distribution of Earth's major climatic zones and factors that affect local and global weather and climate (e.g., deforestation, rainshadow effect, chinook winds, maritime effect)

CLIMATIC ZONES:

The weather in a region is called the **climate** of that region. Unlike the weather, which consists of hourly and daily changes in the atmosphere over a region, climate is the average of all weather conditions in a region over a period of time. Many factors are used to determine the climate of a region including temperature and precipitation. Climate varies from one place to another because of the unequal heating of the Earth's surface. This varied heating of the surface is the result of the unequal distribution of land masses, oceans and polar ice caps.

Climates are classified into three groups: Polar, Tropical, and Temperate. Climates can be affected by the following events: deforestation, global warming, maritime effect, rain shadow effect, and Chinook winds.

FACTORS THAT AFFECT WEATHER/CLIMATE:

The impacts of altitude upon climatic conditions are primarily related to temperature and precipitation. As altitude increases, climatic conditions become increasingly drier and colder. Solar radiation becomes more severe as altitude increases while the effects of convection forces are minimized. Climatic changes as a function of latitude follow a similar pattern (as a reference, latitude moves either north or south from the equator). The climate becomes colder and drier as the distance from the equator increases. Proximity to land or water masses produce climatic conditions based upon the available moisture. Dry and arid climates prevail where moisture is scarce; lush tropical climates can prevail where moisture is abundant. Climate, as described above, depends upon the specific combination of conditions making up an area's environment. Man impacts all environments by producing pollutants in earth, air and water. It follows then that man is a major player in world climatic conditions.

Deforestation is harmful to all aspects of the biosphere. It is immediately harmful to the animals who live there as they lose their homes. Additionally, trees and shrubs assist in biogeochemical cycles. They utilize the elements found in soil and recycle decomposed materials. Trees release oxygen as a process of photosynthesis, and all animals require oxygen to breathe.

The potential effects of global warming are far-reaching. The thickening layer of carbon dioxide pollution threatens to increase average temperatures by 3 to 9 degrees by the end of the century. Global warming potentially will affect weather patterns and increase sea levels. In addition, tropical diseases may also spread into new portions of the globe. Finally, global warming will disrupt ecosystems, causing species extinction and loss of species diversity. The possible solutions for controlling global warming involve the reduction of carbon dioxide in the atmosphere.

Global warming also affects weather patterns and sea levels. In the worst case scenario, melting glaciers will cause sea levels to rise, flooding coastal areas and necessitating the relocation of coastal residents. Higher air temperatures also increase the likelihood of droughts, wildfires, heat waves and intense rainstorms. In addition, higher ocean temperatures may increase the severity of hurricanes and tropical storms. Increased severe weather would negatively affect both local and global economies and cause a rise in heat and weather-related deaths and injuries.

Global warming will also increase the geographic range and virulence of tropical diseases because disease-carrying vectors (e.g. mosquitoes) will spread outside of their normal climatic zones. The spread of disease will have a potentially devastating affect on global health, impacting personal health and increasing the related economic costs of disease treatment and prevention.

Global warming disrupts ecosystems. Rising temperatures will cause the extinction of species that cannot adapt to the climatic changes. Loss of species diversity causes a dangerous imbalance in the ecosystem.

Meteorologists name air masses according to the surface and location over which they originate. Air masses that form over water are called maritime air masses. The effect that they have is known as the maritime effect. The maritime effect refers to the effect that large ocean bodies have on the climate of locations or regions. Maritime climates have a lower range of air temperature than a comparable inland land mass.

The general decrease in precipitation found on the leeward side of a mountain is known as the Rain Shadow Effect. The reduction in precipitation is the result of the warming of descending air caused by compression. The name of winds, specific to North America, also occurring on the leeward side of a mountain, are Chinook winds. This wind is warm and has a low humidity.

Skill 18.6 Identify methods, techniques, tools, and technology used in observing, measuring, and recording weather conditions and in making weather predictions

Every day every one of us is affected by weather regardless if it is the typical thunderstorm with the brief moist air coming down on us from a cumulonimbus cloud or a severe storm with pounding winds that can have wind factors that can cause either hurricanes or twisters called tornados. These are common terms that we can identify with including the term blizzards or ice storms.

The daily newscast relates terms such as dew point and relative humidity and barometric pressure. Suddenly all to common terms become clouded with terms more frequently used by a meteorologist or someone that forecasts weather. **Dew point** is the air temperature at which water vapor begins to condense. **Relative humidity** is the actual amount of water vapor in a certain volume of air compared to the maximum amount of water vapor that this air could hold at a given temperature.

Weather instruments that forecast weather include **aneroid barometer** and the mercury barometer that measures **air pressure**. The air exerts varying pressures on a metal diaphragm that will read air pressure. The mercury barometer operates when atmospheric pressure pushes on a pool of mercury in a glass tube. The higher the pressure the higher up the tube the mercury rises.

Relative humidity is measured by two kinds of weather instruments, the psychrometer and the hair gygrometer. Relative humidity simply indicates the amount of moisture in the air. Relative humidity is defined as a ratio of existing amounts of water vapor and moisture in the air when compared to the maximum amount of moisture that the air can hold at the same given pressure and temperature. Relative humidity is stated as a percentage, so for example the relative humidity can be 100%.

If you were to analyze relative humidity from data, an example might be if a parcel of air is saturated, meaning it now holds all the moisture it can hold at a given temperature, the relative humidity is 100%.

Lesson Plans for teachers to analyze data and predict weather

http://www.srh.weather.gov/srh/jetstream/synoptic/ll_analyze.htm

COMPETENCY 19.0 UNDERSTAND CHARACTERISTICS OF THE HYDROSPHERE AND THE MOVEMENT OF WATER IN THE ENVIRONMENT

Skill 19.1 Demonstrate an understanding of the properties and behavior of water in various states

The change of state of water vapor to a liquid is called condensation. Examples of surface or **ground condensation** include **dew** or **frost**. However, when water vapor condenses in the atmosphere it forms clouds. **Clouds** are composed of water vapor or ice crystals and can contain dust and salts which are known as condensation nuclei. Precipitation is created about these condensation nuclei.

Precipitation includes the following types based on air temperatures and the height and type of air mass from which it will fall and include rain, snow, sleet or hail. Rain falls from high clouds know as nimbostratus, altostratus and cumulonimbus clouds. Snow, sleet and hail are also known as frozen precipitation. The accumulation of ice crystals is snow. Sleet is composed of transparent grains of ice. Hail is created by thunderstorms as ice crystals circulate up and down within the cell gathering bulk until thrown earthward.

Skill 19.2 Identify major categories, characteristics, and distribution of hydrologic systems on Earth (e.g., oceans, glaciers, groundwater, surface waters, water vapor)

The marine biome is the largest biome on earth, covering roughly 75% of the earth. This biome is organized by the depth of the water. The intertidal zone is from the tide line to the edge of the water. The littoral zone is from the waters edge to the open sea. It includes coral reef habitats and is the most densely populated area of the marine biome. The open sea zone is divided into the epipelagic zone and the pelagic zone. The epipelagic zone receives more sunlight and has a larger number of species. The ocean floor is called the benthic zone and is populated with bottom feeders.

The ocean floor has many of the same features that are found on land. The ocean floor has higher mountains than present on land, extensive plains and deeper canyons than present on land. Oceanographers have named different parts of the ocean floor according to their structure. The major parts of the ocean floor are the continental shelf, continental slope, continental rise, abyssal plains, seamounts, Guyots, mid-ocean ridges, and ocean trenches. The deepest trench in the Pacific Ocean is the Marianas Trench, which is about 11 km deep.

A continental glacier covered a large part of North America during the most recent ice age. Evidence of this glacial coverage remains as abrasive grooves, large boulders from northern environments dropped in southerly locations, glacial troughs created by the rounding out of steep valleys by glacial scouring, and the remains of glacial sources called cirques that were created by frost wedging the rock at the bottom of the glacier. Remains of plants and animals found in warm climate have been discovered in the moraines and outwash plains help to support the theory of periods of warmth during the past ice ages.

The Ice Age began about 2 -3 million years ago. This age saw the advancement and retreat of glacial ice over millions of years. Theories relating to the origin of glacial activity include Plate Techtonics where it can be demonstrated that some continental masses, now in temperate climates, were at one time blanketed by ice and snow. Another theory involves changes in the earth's orbit around the sun, changes in the angle of the earth's axis, and the wobbling of the earth's axis. Support for the validity of this theory has come from deep ocean research that indicates a correlation between climatic sensitive microorganisms and the changes in the earth's orbital status.

About 12,000 years ago, a vast sheet of ice covered a large part of the northern United States. This huge, frozen mass had moved southward from the northern regions of Canada as several large bodies of slow-moving ice, or glaciers. A time period in which glaciers advance over a large portion of a continent is called an ice age. A glacier is a large mass of ice that moves or flows over the land in response to gravity. Glaciers form among high mountains and in other cold regions. There are two main types of glaciers: valley glaciers and continental glaciers. Erosion by valley glaciers are characteristic of U-shaped erosion. They produce sharp peaked mountains such as the Matterhorn in Switzerland. Erosion by continental glaciers often ride over mountains in their paths leaving smoothed, rounded mountains and ridges. Two percent of all the available water is fixed and unavailable in ice or the bodies of organisms. Available water includes surface water (lakes, ocean, rivers) and ground water (aquifers, wells). The water located below the surface of the earth is called groundwater. 96% of all available water is from groundwater.

Skill 19.3 Analyze patterns and processes of water circulation through the environment (e.g., water cycle)

Water cycle – Two percent of all the available water is fixed and held in ice or the bodies of organisms. Available water includes surface water (lakes, ocean, and rivers) and ground water (aquifers, wells). Ninety-six percent of all available water is from ground water. Water is recycled through the processes of evaporation and precipitation. The water present now is the water that has been here since our atmosphere formed.

Water that falls to Earth in the form of rain and snow is called **precipitation.** Precipitation is part of a continuous process in which water at the Earth's surface evaporates, condenses into clouds, and returns to Earth. This process is termed the **water cycle**. The water located below the surface is called groundwater.

Skill 19.4 Identify characteristics of watersheds and aquifers and the effects of natural events (e.g., droughts) and human activities (e.g., clear-cutting, pumping of groundwater, building acequias) on watersheds and aquifers

Man has observed (and in some cases created) systems that direct and collect the flow of water. In most situations, water flow runs across land and into small streams that feed larger bodies of water. All of the land that acts like a funnel for water flowing into a single larger body of water is known as a watershed or drainage basin. The watershed includes the streams and rivers that bear the water and the surfaces across which the water runs. Thus, the pollution load and general state of all the land within a watershed has an effect on the health and cleanliness of the body of water to which it drains. Large land features, such as mountains, separate watersheds from one another. However, some portion of water from one watershed may enter the groundwater and ultimately flow towards another adjacent watershed.

Not all water flows to the streams, rivers, and lakes that comprise the above groundwater supply. Some water remains in the soil as groundwater. Additionally, underground rivers are found in areas of karst topography, though these are relatively rare. It is more common for water to collect in underground aquifers. Aquifers are layers of permeable rock or loose material (gravel, sand, or silt) that hold water. Aquifers may be either confined or unconfined. Confined aquifers are deep in the ground and below the water table. Unconfined aquifers border on the water table. The water table is the level at which ground water exists and is always equal to atmospheric pressure. To visualize the entire ground water system, we can imagine a hole dug in wet sand at the beach and a small pool of water within the hole. The wet sand corresponds to the aquifer, the hole to a well or lake, and the level of water in the hole to the water table.

In some cases, people have created reservoirs, artificial storage areas that make large amounts of water readily available. Reservoirs are most often created by damming rivers. A dam is built from cement, soil, or rock and the river fills the newly created reservoir. A reservoir may be created by building a dam either across a valley or around the entire perimeter of an artificial lake (a bunded dam). The former technique is more common and relies on natural features to form a watertight reservoir. However, such a feature must exist to allow this type of construction. A fully bunded dam does not require such a natural feature but does necessitate more construction since a waterproof structure must be built all the way around the reservoir. This structure is typically made from clay and/or cement. Since no river feeds such reservoirs, mechanical pumps are used to fill them from nearby water sources. Occasionally, watertight roofs are added to these reservoirs so they can be used to hold treated water. These are known as service reservoirs.

Both natural events (droughts) and human activities (irrigation, pumping of groundwater) have effects on watersheds. Droughts occur naturally, and are out of human control. Droughts cause riverbeds, streams, and lakes to dry up. Animals perish due to lack of water or move elsewhere to more tolerable conditions (when possible). Even in the best of times, water supply is not always abundant for wildlife because humans reroute water in other, more convenient ways. For example, we capture water and then focus it on our crop areas or lawns. We keep large stores of water in reservoirs where it is useful to us but inaccessible to the rest of nature.

Skill 19.5 Analyze the composition and physical characteristics of oceans (e.g., salinity, density)

Seventy percent of the earth's surface is covered with saltwater which is termed the hydrosphere. The mass of this saltwater is about 1.4 x 10^{24} grams. The ocean waters continuously circulate among different parts of the hydrosphere. There are seven major oceans: the North Atlantic Ocean, South Atlantic Ocean, North Pacific Ocean, South Pacific Ocean, Indian Ocean, Arctic Ocean and the Antarctic Ocean.

Pure water is a combination of the elements hydrogen and oxygen. These two elements make up about 96.5% of ocean water. The remaining portion is made up of dissolved solids. The concentration of these dissolved solids determines the water's salinity.

SALINITY

Salinity is the number of grams of these dissolved salts in 1,000 grams of sea water. The average salinity of ocean water is about 3.5%. In other words, one kilogram of sea water contains about 35 grams of salt. Sodium Chloride or salt (NaCl) is the most abundant of the dissolved salts. The dissolved salts also include smaller quantities of magnesium chloride, magnesium and calcium sulfates, and traces of several other salt elements. Salinity varies throughout the world oceans; the total salinity of the oceans varies from place to place and also varies with depth. Salinity is low near river mouths where the ocean mixes with fresh water and salinity is high in areas of high evaporation rates.

The temperature of the ocean water varies with different latitudes and with ocean depths. Ocean water temperature is about constant to depths of 90 meters (m). The temperature of surface water will drop rapidly from 28° C at the equator to -2° C at the Poles. The freezing point of sea water is lower than the freezing point of pure water. Pure water freezes at 0° C. The dissolved salts in the sea-water keep it at a freezing point of -2° C. The freezing point of seawater may vary depending on its salinity in a particular location.

The ocean can be divided into three temperature zones. The surface layer consists of relatively warm water and exhibits most of the wave action present. The area where the wind and waves churn and mix the water is called the mixed layer. This is the layer where most living creatures are found due to abundant sunlight and warmth. The second layer is called the thermocline and it becomes increasingly cold as its depth increases. This change is due to the lack of energy from sunlight. The layer below the thermocline continues to the deep, dark, very cold and semi-barren ocean floor.

Oozes - the name given to the sediment that contains at least 30% plant or animal shell fragments. Ooze contains calcium carbonate. Deposits that form directly from seawater in the place where they are found are called authigenic deposits. Maganese nodules are authigenic deposits found over large areas of the ocean floor.

DENSITY

Differences in water density also create ocean currents. Water found near the bottom of oceans is the coldest and the densest. Water tends to flow from a denser area to a less dense area. Currents that flow because of a difference in the density of the ocean water are called density currents. Water with a higher salinity is denser than water with a lower salinity. Water that has salinity different from the surrounding water may form a density current.

Skill 19.6 Demonstrate an understanding of the causes and effects of waves and tides, and analyze patterns of ocean circulation (e.g., upwelling, currents) and their effects on weather and climate

The movement of ocean water is caused by the wind, the sun's heat energy, the earth's rotation, the moon's gravitational pull on earth and by underwater earthquakes. Most ocean waves are caused by the impact of winds. Wind blowing over the surface of the ocean transfers energy (friction) to the water and causes waves to form. Waves are also formed by seismic activity on the ocean floor. A wave formed by an earthquake is called a seismic sea wave. These powerful waves can be very destructive with wave heights increasing to 30 m or more near the shore. The crest of a wave is its highest point. The trough of a wave is its lowest point. The distance from wave top to wave top is the wavelength. The wave period is the time between the passing of two successive waves.

World weather patterns are greatly influenced by ocean surface currents in the upper layer of the ocean. These currents continuously move along the ocean surface in specific directions. Ocean currents that flow deep below the surface are called sub-surface currents. These currents are influenced by such factors as the location of land masses in the current's path and the earth's rotation. Surface currents are caused by winds and classified by temperature. Cold currents originate in the Polar regions and flow through surrounding water that is measurably warmer. Those currents with a higher temperature than the surrounding water are called warm currents and can be found near the equator. These currents follow swirling routes around the ocean basins and the equator.

The Gulfstream and the California Current are the two main surface currents that flow along the coastlines of the United States. The Gulfstream is a warm current in the Atlantic Ocean that carries warm water from the equator to the northern parts of the Atlantic Ocean. Benjamin Franklin studied and named the Gulfstream. The California Current is a cold current that originates in the Artic regions and flows southward along the west coast of the United States. Differences in water density also create ocean currents. Water found near the bottom of oceans is the coldest and most dense. Water tends to flow from a denser area to a less dense area. Currents that flow because of a difference in the density of the ocean water are called density currents. Water with a higher salinity is more dense than water with a lower salinity. Water that has a salinity different from the surrounding water may form a density current.

COMPETENCY 20.0 UNDERSTAND INTERACTIONS AMONG EARTH'S SYSTEMS AND CONCEPTS RELATED TO ENVIRONMENTAL SCIENCE

Skill 20.1 Analyze interactions among Earth's lithosphere, hydrosphere, atmosphere, and biosphere and how natural events (e.g., floods, volcanic eruptions) affect and are affected by these interactions

Natural phenomena affect the make up and functioning of ecosystems both directly and indirectly. For example, floods and volcanic eruptions can destroy the fixed portions of an ecosystem, such as plants and microbes. Mobile elements, such as animals, must evacuate or risk injury or death. After a catastrophic event, species of microbes and plants begin to repopulate the ecosystem, beginning a line of secondary succession that eventually leads to the return of higher-level species. Often the area affected by the event returns to something like its original state.

Volcanic eruptions produce large amounts of molten lava and expel large amounts of ash and gas. Molten lava kills and destroys any living organisms it contacts. However, when lava cools and hardens, it provides a rich environment for growth of microbes and plants. Volcanic eruptions also affect ecosystems indirectly. Studies show that the ash and gas released by eruptions can cause a reduction in the area temperature for several years. The volcanic aerosol reflects the sun's rays and creates clouds that have the same effect. In addition, sulfuric acid released by the volcano suppresses the production of greenhouse gases that damage the ozone layer.

Floods destroy microbes and vegetation and kill or force the evacuation of animals. Only when floodwaters recede can an ecosystem begin to return to normal. Floods, however, also have indirect effects. For example, floods can cause permanent soil erosion and nutrient depletion. Such disruptions of the soil can delay and limit an ecosystem's recovery.

Skill 20.2 Identify types of natural resources (e.g., water, wood, land, fossil fuels, minerals) and the consequences of various uses of Earth's natural resources (e.g., global warming, deforestation)

Humans have a tremendous impact on the world's natural resources. The world's natural water supplies are affected by human use. Waterways are major sources for recreation and freight transportation. Oil and wastes from boats and cargo ships pollute the aquatic environment. The aquatic plant and animal life is affected by this contamination. To obtain drinking water, contaminants such as parasites, pollutants and bacteria are removed from raw water through a purification process involving various screening, conditioning and chlorination steps. Most uses of water resources, such as drinking and crop irrigation, require fresh water. Only 2.5% of water on Earth is fresh water, and more than two thirds of this fresh water is frozen in glaciers and polar ice caps.

Consequently, in many parts of the world, water use greatly exceeds supply. This problem is expected to increase in the future.

Plant resources also make up a large part of the world's natural resources. Plant resources are renewable and can be re-grown and restocked. Plant resources can be used by humans to make clothing, buildings and medicines, and can also be directly consumed. Forestry is the study and management of growing forests. This industry provides the wood that is essential for use as construction timber or paper. Cotton is a common plant found on farms of the southern United States. Cotton is used to produce fabric for clothing, sheets and furniture. Another example of a plant resource that is not directly consumed is straw, which is harvested for use in plant growth and farm animal care. The list of plants grown to provide food for the people of the world is extensive. Major crops include corn, potatoes, wheat, sugar, barley, peas, beans, beets, flax, lentils, sunflowers, soybeans, canola and rice. These crops may have alternate uses as well. For example, corn is used to manufacture cornstarch, ethanol fuel, high fructose corn syrup, ink, biodegradable plastics, chemicals used in cosmetics and pharmaceuticals, adhesives and paper products.

Other resources used by humans are known as "non-renewable" resources. Such resources, including fossil fuels, cannot be remade and do not naturally reform at a rate that could sustain human use. Non-renewable resources are therefore depleted and not restored. Presently, non-renewable resources provide the main source of energy for humans. Common fossil fuels used by humans are coal, petroleum and natural gas, which all form from the remains of dead plants and animals through natural processes after millions of years. Because of their high carbon content, when burnt these substances generate high amounts of energy as well as carbon dioxide, which is released back into the atmosphere increasing global warming. To create electricity, energy from the burning of fossil fuels is harnessed to power a rotary engine called a turbine.

Implementation of the use of fossil fuels as an energy source provided for large-scale industrial development.

Mineral resources are concentrations of naturally occurring inorganic elements and compounds located in the earth's crust that are extracted through mining for human use. Minerals have a definite chemical composition and are stable over a range of temperatures and pressures. Construction and manufacturing rely heavily on metals and industrial mineral resources. These metals may include iron, bronze, lead, zinc, nickel, copper or tin. Other industrial minerals are divided into two categories; bulk rocks and ore minerals. Bulk rocks, including limestone, clay, shale and sandstone, are used as aggregate in construction, in ceramics or in concrete. Common ore minerals include calcite, barite and gypsum. Energy from some minerals can be utilized to produce electricity fuel and industrial materials. Mineral resources are also used as fertilizers and pesticides in the industrial context.

Deforestation for urban development has resulted in the extinction or relocation of many species of plants and animals. Animals are forced to leave their forest homes or perish amongst the destruction. The number of plant and animal species that have become extinct due to deforestation is unknown. Scientists have only identified a fraction of the species on Earth. It is known that if the destruction of natural resources continues, there may be no plants or animals successfully reproducing in the wild.

The current energy crisis is largely centered on the uncertain future of fossil fuels. The supplies of fossil fuels are limited and fast declining. Additionally, most oil is now derived from a highly politically volatile area of the world. Finally, continuing to produce energy from fossils fuels is unwise given the damage done by both the disruption to the environment necessary to harvest them and the byproducts of their combustion which causes pollution.

It is important to recognize that a real energy crisis has vast economic implications. Oil, currently the most important fossil fuel, is needed for heating, electricity, and as a raw material for the manufacture of many items, particularly plastics. Additionally, the gasoline made from oil is important in transporting people and goods, including food and other items necessary for life. A disruption in the oil supply often causes rising prices in all sectors and may eventually trigger recession.

Alternative, sustainable energy sources must be found for both economic and ecological reasons.

Skill 20.3 Analyze various strategies for dealing with environmental problems and resource depletion (e.g., mining, nuclear waste, water allocation, agriculture)

As the detrimental effects of pollution have received increasing attention, a number of legislative initiatives have been undertaken to control the pollution and mitigate its effect. The US Environmental Protection Agency (EPA) was established in 1970 to protect human health by reducing damage to the environment. They continue to define and enforce standards for pollution control. The Clean Air (1963) and Clean Water (1977) Acts were two major milestones in the legal control of pollutants.

Various technologies have been invented to reduce the amount of contaminants released. The devices typically use filters to trap the pollutants or rely on chemical reactions to neutralize them. Examples of the former include electrostatic air cleaners and fabric air filters. Examples of the latter include catalytic converters and scrubbers. Materials such as activated charcoal have properties of both types, since they trap and partially neutralize many contaminants, but may still require further chemical treatment. Another key area in which technological advances may be help control pollution is in the search for alternative fuels. Possible alternative energy sources include biodiesel, nuclear power, biomethanol, hydrogen fuel and fuel cells. Increasing the efficiency of solar fuel cells, hydroelectricity, and wind energy may also help reduce dependence on fossil fuels.

Additionally, recycling programs for a variety of materials have begun. Recycling is the reprocessing of materials into new products. Recycling prevents many materials from becoming waste and also avoids the need to harvest new materials. For many materials, recycling requires less energy than virgin production. The most commonly recycled materials are glass, paper, aluminum, asphalt, steel, textiles and plastic. Recycling can occur either pre-consumer or post-consumer.

Finally, many groups now exist to advocate for either conservation or preservation of natural resources. Preservationists are in favor of the strict setting aside of land from human use. Conservationists, on the other hand, prefer sustainable development of land. Preserves are not subjected to any human involvement and so unhealthy overpopulation of a single species is a common problem. Conservation areas or refuges are more actively managed and so typically have more biodiversity. Conservation initiatives are responsible for saving both species and habitats by protecting them from detrimental human involvement.

COMPETENCY 21.0 UNDERSTAND CHARACTERISTICS OF THE SOLAR SYSTEM AND UNIVERSE AND THE METHODS OF EXPLORING THEM

Skill 21.1 Identify characteristics (e.g., mass, temperature, density) and interactions (e.g., gravitational effects) of the major components of the solar system (e.g., the Sun, planets, satellites, asteroids, comets)

The **sun** is considered the nearest star to earth that produces solar energy. By the process of nuclear fusion, hydrogen gas is converted to helium gas. Energy flows out of the core to the surface, then radiation escapes into space.

Parts of the sun include: (1) **core:** the inner portion of the sun where fusion takes place, (2) **photosphere:** considered the surface of the sun which produces **sunspots** (cool, dark areas that can be seen on its surface), (3) **chromosphere:** hydrogen gas causes this portion to be red in color (also found here are solar flares (sudden brightness of the chromosphere) and solar prominences which are gases that shoot outward from the chromosphere), and (4) **corona**, the transparent area of sun visible only during a total eclipse.

There are eight established **planets** in our solar system; Mercury, Venus, Earth, Mars, Jupiter, Saturn, Uranus, and Neptune. Pluto was an established planet in our solar system, but as of Summer 2006, it's status is being reconsidered. The planets are divided into two groups based on distance from the sun. The inner planets include: Mercury, Venus, Earth, and Mars. The outer planets include: Jupiter, Saturn, Uranus, and Neptune.

Mercury -- the closest planet to the sun. Its surface has craters and rocks. The atmosphere is composed of hydrogen, helium and sodium. Mercury was named after the Roman messenger god.

Venus -- has a slow rotation when compared to Earth. Venus and Uranus rotate in opposite directions from the other planets. This opposite rotation is called retrograde rotation. The surface of Venus is not visible due to the extensive cloud cover. The atmosphere is composed mostly of carbon dioxide. Sulfuric acid droplets in the dense cloud cover give Venus a yellow appearance. Venus has a greater greenhouse effect than observed on Earth. The dense clouds combined with carbon dioxide trap heat. Venus was named after the Roman goddess of love.

Earth -- considered a water planet with 70% of its surface covered by water. Gravity holds the masses of water in place. The different temperatures observed on earth allow for the different states (solid. Liquid, gas) of water to exist. The atmosphere is composed mainly of oxygen and nitrogen. Earth is the only planet that is known to support life.

Mars -- the surface of Mars contains numerous craters, active and extinct volcanoes, ridges, and valleys with extremely deep fractures. Iron oxide found in the dusty soil makes the surface seem rust colored and the skies seem pink in color. The atmosphere is composed of carbon dioxide, nitrogen, argon, oxygen and water vapor. Mars has polar regions with ice caps composed of water. Mars has two satellites. Mars was named after the Roman war god.

Jupiter -- largest planet in the solar system. Jupiter has 16 moons. The atmosphere is composed of hydrogen, helium, methane and ammonia. There are white colored bands of clouds indicating rising gas and dark colored bands of clouds indicating descending gases. The gas movement is caused by heat resulting from the energy of Jupiter's core. Jupiter has a Great Red Spot that is thought to be a hurricane type cloud. Jupiter has a strong magnetic field.

Saturn -- the second largest planet in the solar system. Saturn has rings of ice, rock, and dust particles circling it. Saturn's atmosphere is composed of hydrogen, helium, methane, and ammonia. Saturn has 20 plus satellites. Saturn was named after the Roman god of agriculture.

Uranus -- the second largest planet in the solar system with retrograde revolution. Uranus is a gaseous planet. It has 10 dark rings and 15 satellites. Its atmosphere is composed of hydrogen, helium, and methane. Uranus was named after the Greek god of the heavens.

Neptune -- another gaseous planet with an atmosphere consisting of hydrogen, helium, and methane. Neptune has 3 rings and 2 satellites. Neptune was named after the Roman sea god because its atmosphere is the same color as the seas.

Pluto – once considered the smallest planet in the solar system; it's status as a planet is being reconsidered . Pluto's atmosphere probably contains methane, ammonia, and frozen water. Pluto has 1 satellite. Pluto revolves around the sun every 250 years. Pluto was named after the Roman god of the underworld.

COMETS/ASTEROIDS

Comets, asteroids, and meteors.

Astronomers believe that rocky fragments may have been the remains of the birth of the solar system that never formed into a planet. **Asteroids** are found in the region between Mars and Jupiter.

Comets are masses of frozen gases, cosmic dust, and small rocky particles. Astronomers think that most comets originate in a dense comet cloud beyond Pluto. Comets consists of a nucleus, a coma, and a tail. A comet's tail always points away from the sun. The most famous comet, **Halley's Comet,** is named after the person whom first discovered it in 240 B.C. It returns to the skies near earth every 75 to 76 years.

Meteoroids are composed of particles of rock and metal of various sizes. When a meteoroid travels through the earth's atmosphere, friction causes its surface to heat up and it begins to burn. The burning meteoroid falling through the earth's atmosphere is called a **meteor** (also known as a "shooting star").

Meteorites are meteors that strike the earth's surface. A physical example of a meteorite's impact on the earth's surface can be seen in Arizona. The Barringer Crater is a huge meteor crater. There are many other meteor craters throughout the world.

INTERACTIONS:

The mass of any celestial object may be determined by using Newton's laws of motion and his law of gravity.

For example, to determine the mass of the sun, use the following formula:

$$M = \frac{4\pi^2}{G} = \frac{a^3}{P^2}$$

where M = the mass of the sun, G = a constant measured in laboratory experiments, a = the distance of a celestial body in orbit around the Sun from the Sun, and P = the period of the body's orbit.

In our solar system, measurable objects range in mass from the largest, the sun, to the smallest, a near-earth asteroid. (This does not take into account objects with a mass less than 10^{21} kg.)

The surface temperature of an object depends largely upon its proximity to the sun. One exception to this, however, is Venus, which is hotter than Mercury because of its cloud layer that holds heat to the planet's surface. The surface temperatures of the planets range from more than 400 degrees on Mercury and Venus to below -200 degrees on the distant planets.

Most minor bodies in the solar system do not have any atmosphere and, therefore, can easily radiate the heat from the sun. In the case of any celestial object, whether a side is warm or cold depends upon whether it faces the sun or not and the time of rotation. The longer rotation takes, the colder the side facing away from the sun will become, and vice versa.

If the density of an object is less than 1.5 grams per cc, then the object is almost exclusively made of frozen water, ammonia, carbon dioxide, or methane. If the density is less than 1.0, the object must be made of mostly gas. In our solar system, there is only one object with that low a density -- Saturn. If the density is greater than 3.0 grams per cc, then the object is almost exclusively made of rocks and if the density exceeds 5.0 grams per cc, then there must be a nickel-iron core. Densities between 1.5 and 3.0 indicate a rocky-ice mixture.

The density of planets correlates with their distance from the sun. The inner planets (Mercury-Mars) are known as the terrestrial planets because they are rocky, and the outer planets (Jupiter and outward) are known as the icy or Jovian (gaslike) planets.

In order for two bodies to interact gravitationally, they must have significant mass. When two bodies in the solar system interact gravitationally, they orbit about a fixed point (the center of mass of the two bodies). This point lies on an imaginary line between the bodies, joining them such that the distances to each body multiplied by each body's mass are equal. The orbits of these bodies will vary slightly over time because of the gravitational interactions.

Skill 21.2 Analyze the effects of the relative positions, movements, and interactions of the Sun, Earth, and Moon (e.g., seasons, lunar phases, eclipses, tides)

Motion

The earth makes one orbit around the sun in 365.25 days. The earth rotates on its axis once every 24 hours. **Tides** are changes in the level of the ocean caused by the varying gravitational pull of the moon as it orbits the earth. The moon's gravitational force coexists with that of the earth. This interaction produces a common center of gravity between the earth and the moon. This center is called the **barycenter.** As the barycenter rotates around the earth, it causes high tides and low tides. **Neap tides** are low tides that occur twice a month when the sun, earth, and moon are positioned at right angles to one another. **Spring tides** are abnormally high tides that occur twice a month when the sun, earth, and moon are aligned or positioned in a straight line.

Seasons

The **tilt of the earth's axis** allows for the seasonable changes of summer, spring, autumn and winter. As earth revolves around the sun, the angle of the earth's axis changes relative to the sun. This change results in different amounts of sunlight being received by any one spot throughout the year. We recognize these changes as seasonal variations. The **Summer solstice** occurs when the North Pole is tilted toward the sun on June 21 or 22, providing increased daylight hours for the Northern Hemisphere and shorter daylight hours for the Southern Hemisphere. **Winter solstice** occurs when the South Pole is tilted toward the sun on December 21 or 22, providing shorter daylight hours in the Northern Hemisphere and longer daylight hours in the Southern Hemisphere. The **Spring or Vernal Equinox** occurs on March 20 or 21, when the direct energy from the sun falls on the equator providing equal lengths of day and night hours in both hemispheres. **The Autumn Equinox** occurs on September 22 or 23, again providing the equal amounts of day and night hours in both hemispheres.

Eclipses

Eclipses are defined as the passing of one object into the shadow of another object. A **Lunar eclipse** occurs when the moon travels through the shadow of the earth. A **Solar eclipse** occurs when the moon positions itself between the sun and earth.

Phases of the moon

The earth's orientation in respect to the solar system is also responsible for our perception of the phases of the moon. As the earth orbits the sun with a period of 365 days, the moon orbits the earth every 27 days. As the moon circles the earth, its shape in the night sky appears to change. The changes in the appearance of the moon from earth are known as "lunar phases." These phases vary cyclically according to the relative positions of the moon, the earth and the sun. At all times, half of the moon is facing the sun and is thus illuminated by reflecting the sun's light. As the moon orbits the earth and the earth orbits the sun, the half of the moon that faces the sun changes. However, the moon is in synchronous rotation around the earth, meaning that nearly the same side of the moon faces the earth at all times. This side is referred to as the near side of the moon. Lunar phases occur as the earth and moon orbit the sun and the fractional illumination of the moon's near side changes.

When the sun and moon are on opposite sides of the earth, observers on earth perceive a “full moon,” meaning the moon appears circular because the entire illuminated half of the moon is visible. As the moon orbits the earth, the moon “wanes” as the amount of the illuminated half of the moon that is visible from earth decreases. A gibbous moon is between a full moon and a half moon, or between a half moon and a full moon. When the sun and the moon are on the same side of earth, the illuminated half of the moon is facing away from earth, and the moon appears invisible. This lunar phase is known as the “new moon.” The time between each full moon is approximately 29.53 days.

A list of all lunar phases includes:

- New Moon: the moon is invisible or the first signs of a crescent appear
- Waxing Crescent: the right crescent of the moon is visible
- First Quarter: the right quarter of the moon is visible
- Waxing Gibbous: only the left crescent is not illuminated
- Full Moon: the entire illuminated half of the moon is visible
- Waning Gibbous: only the right crescent of the moon is not illuminated
- Last Quarter: the left quarter of the moon is illuminated
- Waning Crescent: only the left crescent of the moon is illuminated

Viewing the moon from the Southern Hemisphere would cause these phases to occur in the opposite order.

Skill 21.3 Demonstrate knowledge of the types and characteristics of celestial objects (e.g., stars, galaxies, black holes, nebulae)

STARS:

Astronomers use groups or patterns of stars called **constellations** as reference points to locate other stars in the sky. Familiar constellations include: Ursa Major (also known as the big bear) and Ursa Minor (known as the little bear). Within the Ursa Major, the smaller constellation, The Big Dipper is found. Within the Ursa Minor, the smaller constellation, The Little Dipper is found.

Different constellations appear as the earth continues its revolution around the sun with the seasonal changes.

Magnitude stars are 21 of the brightest stars that can be seen from earth. These are the first stars noticed at night. In the Northern Hemisphere there are 15 commonly observed first magnitude stars.

GALAXY:

A vast collection of stars are defined as **galaxies**. Galaxies are classified as irregular, elliptical, and spiral. An irregular galaxy has no real structured appearance; most are in their early stages of life. An elliptical galaxy consists of smooth ellipses, containing little dust and gas, but composed of millions or trillions of stars. Spiral galaxies are disk-shaped and have extending arms that rotate around its dense center. Earth's galaxy is found in the Milky Way and it is a spiral galaxy.

Terms related to deep space

A **pulsar** is defined as a variable radio source that emits signals in very short, regular bursts; believed to be a rotating neutron star.

A **quasar** is defined as an object that photographs like a star but has an extremely large redshift and a variable energy output; believed to be the active core of a very distant galaxy.

BLACK HOLE:

Black holes are defined as an object that has collapsed to such a degree that light can not escape from its surface; light is trapped by the intense gravitational field.

NEBULAE:

Nebulae are clouds of dust and gas. These clouds are the birth place for stars. Nebulae are formed either because of a nearby supernova explosion, or because a star has died, expelling some of its outermost content.

Skill 21.4 Demonstrate an understanding of various scientific theories of the origin of the universe (e.g., Big Bang)

Two main hypotheses of the origin of the solar system are: (1) **the tidal hypothesis** and (2) **the condensation hypothesis**.

The tidal hypothesis proposes that the solar system began with a near collision of the sun and a large star. Some astronomers believe that as these two stars passed each other, the great gravitational pull of the large star extracted hot gases out of the sun. The mass from the hot gases started to orbit the sun, which began to cool then condensing into the nine planets. (Few astronomers support this example).

The condensation hypothesis proposes that the solar system began with rotating clouds of dust and gas. Condensation occurred in the center forming the sun and the smaller parts of the cloud formed the nine planets. (This example is widely accepted by many astronomers).

Two main theories to explain the origins of the universe include: (1) **The Big Bang Theory** and (2) **The Steady-State Theory.**

The Big Bang Theory has been widely accepted by many astronomers. It states that the universe originated from a magnificent explosion spreading mass, matter and energy into space. The galaxies formed from this material as it cooled during the next half-billion years.

The Steady-State Theory is the least accepted theory. It states that the universe is a continuously being renewed. Galaxies move outward and new galaxies replace the older galaxies. Astronomers have not found any evidence to prove this theory.

The future of the universe is hypothesized with the Oscillating Universe Hypothesis. It states that the universe will oscillate or expand and contract. Galaxies will move away from one another and will in time slow down and stop. Then a gradual moving toward each other will again activate the explosion or The Big Bang theory.

Skill 21.5 Interpret simple data derived from remote and direct observations of the solar system and universe (e.g., evidence of planets near stars other than the Sun)

Spectral analysis

The **spectroscope** is a device or an attachment for telescopes that is used to separate white light into a series of different colors by wave lengths. This series of colors of light is called a **spectrum**. A **spectrograph** can photograph a spectrum. Wavelengths of light have distinctive colors. The color red has the longest wavelength and violet has the shortest wavelength. Wavelengths are arranged to form an **electromagnetic spectrum**. They range from very long radio waves to very short gamma rays. Visible light covers a small portion of the electromagnetic spectrum. Spectroscopes observe the spectra, temperatures, pressures, and also the movement of stars. The movements of stars indicate if they are moving toward, or away from earth.

If a star is moving towards earth, light waves compress and the wavelengths of light seem shorter. This will cause the entire spectrum to move towards the blue or violet end of the spectrum.

If a star is moving away from earth, light waves expand and the wavelengths of light seem longer. This will cause the entire spectrum to move towards the red end of the spectrum.

Skill 21.6 Demonstrate knowledge of human space exploration, methods and technology used to gather data about space (e.g., types of telescopes, space probes), and benefits to society of technological advances associated with space exploration

KNOWLEDGE:

Knowledge of astronomical measurement.

The three formulas astronomers use for calculating distances in space are: (1) the **AU or astronomical unit**, (2) **the LY or Light year,** and (3) **the parsec**. It is important to remember that these formulas are measured in distances not time.

The distance between the earth and the sun is about 150×10^{6} km. This distance is known as an astronomical unit or AU. This formula is used to measure distances within the solar system, it is not used to measure time.

The distance light travels in one year is a light year (9.5×10^{12} km). This formula is used to measure distances in space, it does not measure time. Large distances are measured in parsecs. One parsec equals 3.26 light-years. There are approximately 63,000 AU's in one light year or,
9.5×10^{12} km/ 150×10^{6} km $= 6.3 \times 10^{4}$ AU

TELESCOPES:

Knowledge of telescope types

Galileo was the first person to use telescopes to observe the solar system. He invented the first refracting telescope. A **refracting telescope** uses lenses to bend light rays to focus the image.

Sir Isaac Newton invented the **reflecting telescope** using mirrors to gather light rays on a curved mirror which produces a small focused image.

The world's largest telescope is located in Mauna Kea, Hawaii. It uses multiple mirrors to gather light rays.

The **Hubble Space telescope** uses a **single-reflector mirror**. It provides an opportunity for astronomers to observe objects seven times further away. Even those objects that are 50 times fainter can be viewed better than by any telescope on earth. There are future plans to make repairs and install new mirrors and other equipment on the Hubble Space telescope.

Refracting and reflecting telescopes are considered **optical telescopes** since they gather visible light and focus it to produce images. A different type of telescope that collects invisible radio waves created by the sun and stars is called a **radio telescope.**

Radio telescopes consists of a reflector or dish with special receivers. The reflector collects radio waves that are created by the sun and stars. Using a radio telescope has many advantages. They can receive signals 24 hours a day, can operate in any kind of weather and dust particles or clouds do not interfere with its performance. The most impressive aspect of the radio telescope is its ability to detect objects from such great distances in space.

The world's largest radio telescope is located in Arecibo, Puerto Rico. It has a collecting dish antenna of more than 300 meters in diameter.

SPACE EXPLORATION:

Though outer space has been a subject of fascination throughout human history, space exploration refers particularly to the travel into outer space to discover new features and facts. Though space exploration continues today, it was at its height in the late 20th century, when much progress was made over just a few years.

Some of earliest forays into true space exploration were unmanned missions involving space probes. The probes are controlled remotely from Earth and have been shot into outer space and immediately returned, placed into orbit around our planet, and sent to and past the other planets in our solar system. The first was the USSR's Sputnik I in October of 1957. It was the first man-made object ever launched into space. This was the beginning of the "space race" between the USSR and USA. The USA's first successful launch of a space probe occurred with Vanguard I in December of 1957. A few early, unmanned missions were space probes carrying animals, such as the Soviet dog Laika that became the first animal in orbit in November of 1957. Animals are included only for research purposes in current missions. Space probes are still used for certain applications where risk, cost, or duration makes manned missions impractical. The Voyager probes are among the most famous probes. They were launched to take advantage of the favorable planetary alignment in the late 1970s. They returned data and fascinating pictures from Jupiter and Saturn as well as information from beyond our solar system. It is hoped that, as technology continues to improve, space probes will be allow us to investigate space even farther away from Earth.

The first manned mission occurred in 1961, when the USSR launched Yuri Gagarin aboard Vostok I into space. A year later, the American John Glenn became the first man to orbit the Earth. The USA finally pulled well ahead in the space race in 1969, when Neil Armstrong and Buzz Aldrin became the first men to reach the moon aboard Apollo 11. Reusable space shuttles were a large step forward in allowing manned missions. The first space shuttle to enter outer space was the Columbia, though other famous US shuttles include the Challenger, Atlantis and Endeavour. Shuttles are now used to conduct experiments and to transport astronauts to and from space stations.

Space stations now serve as key tool in space exploration. A space station is any artificial structure designed to house, but not transport, humans living in space. The first space station was Salyut 1, launched by the USSR in 1971. This, like all space stations up to the present, was a low earth orbital station. Other space stations include Skylab, Salyuts 2-7, Mir, and the International Space Station. Only the International Space Station is currently in use. Space stations offer an excellent environment to run long-term experiments in outer space. However, they are not suitable for human life beyond a few months because of the low gravity, high radiation, and other less understood factors. Much progress needs to be made before human beings will be able to live permanently in space. In fact, the future of manned missions is somewhat uncertain, as there is some debate about how necessary they are. Many speculate that significant cost and risk could be avoided with the use of robots. Currently, humans in space perform many experiments and conduct necessary repairs on equipment.

BENEFITS:

Space exploration, like all scientific endeavors, provides the expansion of our knowledge about how the universe works. However, given the relatively high cost of space exploration, further justification is needed. Firstly, money spent on space research creates many jobs and so has economic benefits. Secondly, as space exploration has become an increasingly international affair, it has served to increase cooperation between nations and generate goodwill. Note that such cooperation also decreases the financial burden for individual countries. However, one of the greatest benefits of space exploration is the potential for the transfer of technology. A vast array of technologies developed to further space exploration have found broader applications. These include communication devices, satellite operations, electronics, fabrics, and other materials. For instance, the technology used in smoke detectors was developed for NASA's Skylab spacecraft in the 1970s and quartz timing crystals used in nearly all wristwatches were developed as timing devices for the Apollo lunar missions.

Sample Test

DIRECTIONS: Read each item and select the correct response. The answer key follows.

1. After an experiment, the scientist states that s/he believes a change in the color of a liquid is due to a change of pH. This is an example of __________ .

A. observing.

B. inferring.

C. measuring.

D. classifying.

2. When is a hypothesis formed?

A. Before the data is taken.

B. After the data is taken.

C. After the data is analyzed.

D. Concurrent with graphing the data.

3. Who determines the laws regarding the use of safety glasses in the classroom?

A. The state.

B. The school site.

C. The Federal government.

D. The district level.

4. If one inch equals 2.54 cm how many mm in 1.5 feet? (APPROXIMATELY)

A. 18 mm.

B. 1800 mm.

C. 460 mm.

D. 4,600 mm.

5. Which of the following instruments measures wind speed?

A. A barometer.

B. An anemometer.

C. A thermometer.

D. A weather vane.

6. Sonar works by __________ .

A. timing how long it takes sound to reach a certain speed.

B. bouncing sound waves between two metal plates.

C. bouncing sound waves off an underwater object and timing how long it takes for the sound to return.

D. evaluating the motion and amplitude of sound.

7. The measure of the pull of the earth's gravity on an object is called __________ .

A. mass number.

B. atomic number.

C. mass.

D. weight.

8. Which reaction below is a decomposition reaction?

A. $HCl + NaOH \rightarrow NaCl + H_2O$

B. $C + O_2 \rightarrow CO_2$

C. $2H_2O \rightarrow 2H_2 + O_2$

D. $CuSO_4 + Fe \rightarrow FeSO_4 + Cu$

9. The Law of Conservation of Energy states that _________.

A. There must be the same number of products and reactants in any chemical equation.

B. Objects always fall toward large masses such as planets.

C. Energy is neither created nor destroyed, but may change form.

D. Lights must be turned off when not in use, by state regulation.

10. Which parts of an atom are located inside the nucleus?

A. electrons and neutrons.

B. protons and neutrons.

C. protons only.

D. neutrons only.

11. The elements in the modern Periodic Table are arranged __________ .

A. in numerical order by atomic number.

B. randomly.

C. in alphabetical order by chemical symbol.

D. in numerical order by atomic mass.

12. Carbon bonds with hydrogen by __________ .

A. ionic bonding.

B. non-polar covalent bonding.

C. polar covalent bonding.

D. strong nuclear force.

13. Vinegar is an example of a __________ .

A. strong acid.

B. strong base.

C. weak acid.

D. weak base.

14. Which of the following is not a nucleotide?

A. adenine.

B. alanine.

C. cytosine.

D. guanine.

15. When measuring the volume of water in a graduated cylinder, where does one read the measurement?

A. At the highest point of the liquid.

B. At the bottom of the meniscus curve.

C. At the closest mark to the top of the liquid.

D. At the top of the plastic safety ring.

16. A duck's webbed feet are examples of __________ .

A. mimicry.

B. structural adaptation.

C. protective resemblance.

D. protective coloration.

17. What cell organelle contains the cell's stored food?

A. Vacuoles.

B. Golgi Apparatus.

C. Ribosomes.

D. Lysosomes.

18. The first stage of mitosis is called __________ .

A. telophase.

B. anaphase.

C. prophase.

D. mitophase.

19. The Doppler Effect is associated most closely with which property of waves?

A. amplitude.

B. wavelength.

C. frequency.

D. intensity.

20. Viruses are responsible for many human diseases including all of the following eXCEPT _________ ?

A. influenza.

B. A.I.D.S.

C. the common cold.

D. strep throat.

21. A series of experiments on pea plants formed by ________ showed that two invisible markers existed for each trait, and one marker dominated the other.

A. Pasteur.

B. Watson and Crick.

C. Mendel.

D. Mendeleev.

22. Formaldehyde should not be used in school laboratories for the following reason:

A. it smells unpleasant.

B. it is a known carcinogen.

C. it is expensive to obtain.

D. it is explosive.

23. Amino acids are carried to the ribosome in protein synthesis by _________ .

A. transfer RNA (tRNA).

B. messenger RNA (mRNA).

C. ribosomal RNA (rRNA).

D. transformation RNA (trRNA).

24. When designing a scientific experiment, a student considers all the factors that may influence the results. The process goal is to ________.

A. recognize and manipulate independent variables.

B. recognize and record independent variables.

C. recognize and manipulate dependent variables.

D. recognize and record dependent variables.

25. Since ancient times, people have been entranced with bird flight. What is the key to bird flight?

A. Bird wings are a particular shape and composition.

B. Birds flap their wings quickly enough to propel themselves.

C. Birds take advantage of tailwinds.

D. Birds take advantage of crosswinds.

26. Laboratory researchers have classified fungi as distinct from plants because the cell walls of fungi __________ .

A. contain chitin.

B. contain yeast.

C. are more solid.

D. are less solid.

27. In a fission reactor, "heavy water" is used to __________ .

A. terminate fission reactions.

B. slow down neutrons and moderate reactions.

C. rehydrate the chemicals.

D. initiate a chain reaction.

28. The transfer of heat by electromagnetic waves is called __________ .

A. conduction.

B. convection.

C. phase change.

D. radiation.

29. When heat is added to most solids, they expand. Why is this the case?

A. The molecules get bigger.

B. The faster molecular motion leads to greater distance between the molecules.

C. The molecules develop greater repelling electric forces.

D. The molecules form a more rigid structure.

30. The force of gravity on earth causes all bodies in free fall to __________ .

A. fall at the same speed.

B. accelerate at the same rate.

C. reach the same terminal velocity.

D. move in the same direction.

31. Sound waves are produced by __________ .

A. pitch.

B. noise.

C. vibrations.

D. sonar.

32. Resistance is measured in units called __________ .

A. watts.

B. volts.

C. ohms.

D. current.

33. Sound can be transmitted in all of the following EXCEPT __________ .

A. air.

B. water.

C. A diamond.

D. a vacuum.

34. As a train approaches, the whistle sounds __________ .

A. higher, because it has a higher apparent frequency.

B. lower, because it has a lower apparent frequency.

C. higher, because it has a lower apparent frequency.

D. lower, because it has a higher apparent frequency.

35. The speed of light is different in different materials. This is responsible for __________ .

A. interference.

B. refraction.

C. reflection.

D. relativity.

36. A converging lens produces a real image ________________.

A. always.

B. never.

C. when the object is within one focal length of the lens.

D. when the object is further than one focal length from the lens.

37. The electromagnetic radiation with the longest wave length is/are ___________.

A. radio waves.

B. red light.

C. X-rays.

D. ultraviolet light.

38. Under a 440 power microscope, an object with diameter 0.1 millimeter appears to have a diameter of _________ .

A. 4.4 millimeters.

B. 44 millimeters.

C. 440 millimeters.

D. 4400 millimeters.

39. Separating blood into blood cells and plasma involves the process of _________ .

A. electrophoresis.

B. centrifugation.

C. spectrophotometry.

D. chromatography.

40. Experiments may be done with any of the following animals except _________ .

A. birds.

B. invertebrates.

C. lower order life.

D. frogs.

41. For her first project of the year, a student is designing a science experiment to test the effects of light and water on plant growth. You should recommend that she ________________.

A. manipulate the temperature also.

B. manipulate the water pH also.

C. determine the relationship between light and water unrelated to plant growth.

D. omit either water or light as a variable.

42. In a laboratory report, what is the abstract?

A. The abstract is a summary of the report, and is the first section of the report.

B. The abstract is a summary of the report, and is the last section of the report.

C. The abstract is predictions for future experiments, and is the first section of the report.

D. The abstract is predictions for future experiments, and is the last section of the report.

43. What is the scientific method?

A. It is the process of doing an experiment and writing a laboratory report.

B. It is the process of using open inquiry and repeatable results to establish theories.

C. It is the process of reinforcing scientific principles by confirming results.

D. It is the process of recording data and observations.

44.Identify the control in the following experiment: A student had four corn plants and was measuring photosynthetic rate (by measuring growth mass). Half of the plants were exposed to full (constant) sunlight, and the other half were kept in 50% (constant) sunlight.

A. The control is a set of plants grown in full (constant) sunlight.

B. The control is a set of plants grown in 50% (constant) sunlight.

C. The control is a set of plants grown in the dark.

D. The control is a set of plants grown in a mixture of natural levels of sunlight.

45. In an experiment measuring the growth of bacteria at different temperatures, what is the independent variable?

A. Number of bacteria.

B. Growth rate of bacteria.

C. Temperature.

D. Light intensity.

46. A scientific law__________.

A. proves scientific accuracy.

B. may never be broken.

C. may be revised in light of new data.

D. is the result of one excellent experiment.

47. Which is the correct order of methodology?

1. **collecting data**
2. **planning a controlled experiment**
3. **drawing a conclusion**
4. **hypothesizing a result**
5. **re-visiting a hypothesis to answer a question**

A. 1,2,3,4,5

B. 4,2,1,3,5

C. 4,5,1,3,2

D. 1,3,4,5,2

48. Which is the most desirable tool to use to heat substances in a middle school laboratory?

A. Alcohol burner.

B. Freestanding gas burner.

C. Bunsen burner.

D. Hot plate.

49. Newton's Laws are taught in science classes because __________.

A. they are the correct analysis of inertia, gravity, and forces.

B. they are a close approximation to correct physics, for usual Earth conditions.

C. they accurately incorporate relativity into studies of forces.

D. Newton was a well-respected scientist in his time.

50. Which of the following is most accurate?

A. Mass is always constant; Weight may vary by location.

B. Mass and Weight are both always constant.

C. Weight is always constant; Mass may vary by location.

D. Mass and Weight may both vary by location.

51. Chemicals should be stored

A. in the principal's office.

B. in a dark room.

C. in an off-site research facility.

D. according to their reactivity with other substances.

52. Which of the following is the worst choice for a school laboratory activity?

A. A genetics experiment tracking the fur color of mice.

B. Dissection of a preserved fetal pig.

C. Measurement of goldfish respiration rate at different temperatures.

D. Pithing a frog to watch the circulatory system.

53. Who should be notified in the case of a serious chemical spill?

A. The custodian.

B. The fire department or their municipal authority.

C. The science department chair.

D. The School Board.

54. A scientist exposes mice to cigarette smoke, and notes that their lungs develop tumors. Mice that were not exposed to the smoke do not develop as many tumors. Which of the following conclusions may be drawn from these results?

I. Cigarette smoke causes lung tumors.
II. Cigarette smoke exposure has a positive correlation with lung tumors in mice.
III. Some mice are predisposed to develop lung tumors.
IV. Mice are often a good model for humans in scientific research.

A. I and II only.

B. II only.

C. I , II, and III only.

D. II and IV only.

55. In which situation would a science teacher be legally liable?

A The teacher leaves the classroom for a telephone call and a student slips and injures him/herself.

B. A student removes his/her goggles and gets acid in his/her eye.

C. A faulty gas line in the classroom causes a fire.

D. A student cuts him/herself with a dissection scalpel.

56. Which of these is the best example of 'negligence'?

A. A teacher fails to give oral instructions to those with reading disabilities.

B. A teacher fails to exercise ordinary care to ensure safety in the classroom.

C. A teacher displays inability to supervise a large group of students.

D. A teacher reasonably anticipates that an event may occur, and plans accordingly.

57. Which item should always be used when handling glassware?

A. Tongs.

B. Safety goggles.

C. Gloves.

D. Buret stand.

58. Which of the following is NOT a necessary characteristic of living things?

A. Movement.

B. Reduction of local entropy.

C. Ability to cause change in local energy form.

D. Reproduction.

59. What are the most significant and prevalent elements in the biosphere?

A. Carbon, Hydrogen, Oxygen, Nitrogen, Phosphorus.

B. Carbon, Hydrogen, Sodium, Iron, Calcium.

C. Carbon, Oxygen, Sulfur, Manganese, Iron.

D. Carbon, Hydrogen, Oxygen, Nickel, Sodium, Nitrogen.

60. All of the following measure energy EXCEPT for __________

A. joules.

B. calories.

C. watts.

D. ergs.

61. Identify the correct sequence of organization of living things from lower to higher order:

A. Cell, Organelle, Organ, Tissue, System, Organism.

B. Cell, Tissue, Organ, Organelle, System, Organism.

C. Organelle, Cell, Tissue, Organ, System, Organism.

D. Organelle, Tissue, Cell, Organ, System, Organism.

62. Which kingdom is comprised of organisms made of one cell with no nuclear membrane?

A. Monera.

B. Protista.

C. Fungi.

D. Algae.

63. Which of the following is found in the least abundance in organic molecules?

A. Phosphorus.

B. Potassium.

C. Carbon.

D. Oxygen.

64. Catalysts assist reactions by __________ .

A. lowering effective activation energy.

B. maintaining precise pH levels.

C. keeping systems at equilibrium.

D. adjusting reaction speed.

65. Accepted procedures for preparing solutions should be made with __________ .

A. alcohol.

B. hydrochloric acid.

C. distilled water.

D. tap water.

66. Enzymes speed up reactions by __________ .

A. utilizing ATP.

B. lowering pH, allowing reaction speed to increase.

C. increasing volume of substrate.

D. lowering energy of activation.

67. When you step out of the shower, the floor feels colder on your feet than the bathmat. Which of the following is the correct explanation for this phenomenon?

A. The floor is colder than the bathmat.

B. Your feet have a chemical reaction with the floor, but not the bathmat.

C. Heat is conducted more easily into the floor.

D. Water is absorbed from your feet into the bathmat.

68. Which of the following is NOT considered ethical behavior for a scientist?

A. Using unpublished data and citing the source.

B. Publishing data before other scientists have had a chance to replicate results.

C. Collaborating with other scientists from different laboratories.

D. Publishing work with an incomplete list of citations.

69. The chemical equation for water formation is: $2H_2 + O_2 \rightarrow 2H_2O$. Which of the following is an INCORRECT interpretation of this equation?

A. Two moles of hydrogen gas and one mole of oxygen gas combine to make two moles of water.

B. Two grams of hydrogen gas and one gram of oxygen gas combine to make two grams of water.

C. Two molecules of hydrogen gas and one molecule of oxygen gas combine to make two molecules of water.

D. Four atoms of hydrogen (combined as a diatomic gas) and two atoms of oxygen (combined as a diatomic gas) combine to make two molecules of water.

70. Energy is measured with the same units as _____________.

A. force.

B. momentum.

C. work.

D. power.

71. If the volume of a confined gas is increased, what happens to the pressure of the gas? You may assume that the gas behaves ideally, and that temperature and number of gas molecules remain constant.

A. The pressure increases.

B. The pressure decreases.

C. The pressure stays the same.

D. There is not enough information given to answer this question.

72. A product of anaerobic respiration in animals is __________.

A. carbon dioxide.

B. lactic acid.

C. oxygen.

D. sodium chloride

73. A Newton is fundamentally a measure of __________ .

A. force.

B. momentum.

C. energy.

D. gravity.

74. Which change does NOT affect enzyme rate?

A. Increase the temperature.

B. Add more substrate.

C. Adjust the pH.

D. Use a larger cell.

75. Which of the following types of rock are made from magma?

A. Fossils

B. Sedimentary

C. Metamorphic

D. Igneous

76. Which of the following is NOT an acceptable way for a student to acknowledge sources in a laboratory report?

A. The student tells his/her teacher what sources s/he used to write the report.

B. The student uses footnotes in the text, with sources cited, but not in correct MLA format.

C. The student uses endnotes in the text, with sources cited, in correct MLA format.

D. The student attaches a separate bibliography, noting each use of sources.

77. Animals with a notochord or backbone are in the phylum

A. Arthropoda.

B. Chordata.

C. Mollusca.

D. Mammalia.

78. Which of the following is a correct explanation for scientific 'evolution'?

A. Giraffes need to reach higher for leaves to eat, so their necks stretch. The giraffe babies are then born with longer necks. Eventually, there are more long-necked giraffes in the population.

B. Giraffes with longer necks are able to reach more leaves, so they eat more and have more babies than other giraffes. Eventually, there are more long-necked giraffes in the population.

C. Giraffes want to reach higher for leaves to eat, so they release enzymes into their bloodstream, which in turn causes fetal development of longer-necked giraffes. Eventually, there are more long-necked giraffes in the population.

D. Giraffes with long necks are more attractive to other giraffes, so they get the best mating partners and have more babies. Eventually, there are more long-necked giraffes in the population.

79. Which of the following is a correct definition for 'chemical equilibrium'?

A. Chemical equilibrium is when the forward and backward reaction rates are equal. The reaction may continue to proceed forward and backward.

B. Chemical equilibrium is when the forward and backward reaction rates are equal, and equal to zero. The reaction does not continue.

C. Chemical equilibrium is when there are equal quantities of reactants and products.

D. Chemical equilibrium is when acids and bases neutralize each other fully.

80. Which of the following data sets is properly represented by a bar graph?

A. Number of people choosing to buy cars, vs. Color of car bought.

B. Number of people choosing to buy cars, vs. Age of car customer.

C. Number of people choosing to buy cars, vs. Distance from car lot to customer home.

D. Number of people choosing to buy cars, vs. Time since last car purchase.

81. In a science experiment, a student needs to dispense very small measured amounts of liquid into a well-mixed solution. Which of the following is the best choice for his/her equipment to use?

A. Buret with Buret Stand, Stir-plate, Stirring Rod, Beaker.

B. Buret with Buret Stand, Stir-plate, Beaker.

C. Volumetric Flask, Dropper, Graduated Cylinder, Stirring Rod.

D. Beaker, Graduated Cylinder, Stir-plate.

82. A laboratory balance is most appropriately used to measure the mass of which of the following?

A. Seven paper clips.

B. Three oranges.

C. Two hundred cells.

D. One student's elbow.

83. All of the following are measured in units of length, EXCEPT for:

A. Perimeter.

B. Distance.

C. Radius.

D. Area.

84. What is specific gravity?

A. The mass of an object.

B. The ratio of the density of a substance to the density of water.

C. Density.

D. The pull of the earth's gravity on an object.

85. What is the most accurate description of the Water Cycle?

A. Rain comes from clouds, filling the ocean. The water then evaporates and becomes clouds again.

B. Water circulates from rivers into groundwater and back, while water vapor circulates in the atmosphere.

C. Water is conserved except for chemical or nuclear reactions, and any drop of water could circulate through clouds, rain, ground-water, and surface-water.

D. Weather systems cause chemical reactions to break water into its atoms.

86. The scientific name CANIS FAMILIARIS refers to the animal's __________.

A. kingdom and phylum.

B. genus and species.

C. class and species.

D. type and family.

87. Members of the same animal species __________ .

A. look identical.

B. never adapt differently.

C. are able to reproduce with one another.

D. are found in the same location.

88. Which of the following is/are true about scientists?

- **I. Scientists usually work alone.**
- **II. Scientists usually work with other scientists.**
- **III. Scientists achieve more prestige from new discoveries than from replicating established results.**
- **IV. Scientists keep records of their own work, but do not publish it for outside review.**

A. I and IV only.

B. II only.

C. II and III only.

D. I and IV only.

89. What is necessary for ion diffusion to occur spontaneously?

A. Carrier proteins.

B. Energy from an outside source.

C. A concentration gradient.

D. Cell flagellae.

90. All of the following are considered Newton's Laws EXCEPT for:

A. An object in motion will continue in motion unless acted upon by an outside force.

B. For every action force, there is an equal and opposite reaction force.

C. Nature abhors a vacuum.

D. Mass can be considered the ratio of force to acceleration.

91. A cup of hot liquid and a cup of cold liquid are both sitting in a room at comfortable room temperature and humidity. Both cups are thin plastic. Which of the following is a true statement?

A. There will be fog on the outside of the hot liquid cup, and also fog on the outside of the cold liquid cup.

B. There will be fog on the outside of the hot liquid cup, but not on the cold liquid cup.

C. There will be fog on the outside of the cold liquid cup, but not on the hot liquid cup.

D. There will not be fog on the outside of either cup.

92. A ball rolls down a smooth hill. You may ignore air resistance. Which of the following is a true statement?

A. The ball has more energy at the start of its descent than just before it hits the bottom of the hill, because it is higher up at the beginning.

B. The ball has less energy at the start of its descent than just before it hits the bottom of the hill, because it is moving more quickly at the end.

C. The ball has the same energy throughout its descent, because positional energy is converted to energy of motion.

D. The ball has the same energy throughout its descent, because a single object (such as a ball) cannot gain or lose energy.

93. A long silver bar has a temperature of 50 degrees Celsius at one end and 0 degrees Celsius at the other end. The bar will reach thermal equilibrium (barring outside influence) by the process of heat _______________.

A. conduction.

B. convection.

C. radiation.

D. phase change.

94. ____________ are cracks in the plates of the earth's crust, along which the plates move.

A. Faults.

B. Ridges.

C. Earthquakes.

D. Volcanoes.

95. Fossils are usually found in _____________ rock.

A. igneous.

B. sedimentary.

C. metamorphic.

D. cumulus.

96. Which of the following is NOT a common type of acid in 'acid rain' or acidified surface water?

A. Nitric acid.

B. Sulfuric acid.

C. Carbonic acid.

D. Hydrofluoric acid.

97. Which of the following is NOT true about phase change in matter?

A. Solid water and liquid ice can coexist at water's freezing point.

B. At 7 degrees Celsius, water is always in liquid phase.

C. Matter changes phase when enough energy is gained or lost.

D. Different phases of matter are characterized by differences in molecular motion.

98. Which of the following is the longest (largest) unit of geological time?

A. Solar Year.

B. Epoch.

C. Period.

D. Era.

99. Extensive use of antibacterial soap has been found to increase the virulence of certain infections in hospitals. Which of the following might be an explanation for this phenomenon?

A. Antibacterial soaps do not kill viruses.

B. Antibacterial soaps do not incorporate the same antibiotics used as medicine.

C. Antibacterial soaps kill a lot of bacteria, and only the hardiest ones survive to reproduce.

D. Antibacterial soaps can be very drying to the skin.

100. Which of the following is a correct explanation for astronaut 'weightlessness'?

A. Astronauts continue to feel the pull of gravity in space, but they are so far from planets that the force is small.

B. Astronauts continue to feel the pull of gravity in space, but spacecraft have such powerful engines that those forces dominate, reducing effective weight.

C. Astronauts do not feel the pull of gravity in space, because space is a vacuum.

D. Astronauts do not feel the pull of gravity in space, because black hole forces dominate the force field, reducing their masses.

101. The theory of 'sea floor spreading' explains _________.

A. the shapes of the continents.

B. how continents collide.

C. how continents move apart.

D. how continents sink to become part of the ocean floor.

102. Which of the following animals are most likely to live in a tropical rain forest?

A. Reindeer.

B. Monkeys.

C. Puffins.

D. Bears.

103. Which of the following is NOT a type of volcano?

A. Shield Volcanoes.

B. Composite Volcanoes.

C. Stratus Volcanoes.

D. Cinder Cone Volcanoes.

104. Which of the following is NOT a property of metalloids?

A. Metalloids are solids at standard temperature and pressure.

B. Metalloids can conduct electricity to a limited extent.

C. Metalloids are found in groups 13 through 17.

D. Metalloids all favor ionic bonding.

105. Which of these is a true statement about loamy soil?

A. Loamy soil is gritty and porous.

B. Loamy soil is smooth and a good barrier to water.

C. Loamy soil is hostile to microorganisms.

D. Loamy soil is velvety and clumpy.

106. Lithification refers to the process by which unconsolidated sediments are transformed into ___________.

A. metamorphic rocks.

B. sedimentary rocks.

C. igneous rocks.

D. lithium oxide.

107. Igneous rocks can be classified according to which of the following?

A. Texture.

B. Composition.

C. Formation process.

D. All of the above.

108. Which of the following is the most accurate definition of a non-renewable resource?

A. A nonrenewable resource is never replaced once used.

B. A nonrenewable resource is replaced on a timescale that is very long relative to human life-spans.

C. A nonrenewable resource is a resource that can only be manufactured by humans.

D. A nonrenewable resource is a species that has already become extinct.

109. The theory of 'continental drift' is supported by which of the following?

A. The way the shapes of South America and Europe fit together.

B. The way the shapes of Europe and Asia fit together.

C. The way the shapes of South America and Africa fit together.

D. The way the shapes of North America and Antarctica fit together.

110. When water falls to a cave floor and evaporates, it may deposit calcium carbonate. This process leads to the formation of which of the following?

A. Stalactites.

B. Stalagmites.

C. Fault lines.

D. Sedimentary rocks.

111. A child has type O blood. Her father has type A blood, and her mother has type B blood. What are the genotypes of the father and mother, respectively?

A. AO and BO.

B. AA and AB.

C. OO and BO.

D. AO and BB.

112. Which of the following is the best definition for 'meteorite'?

A. A meteorite is a mineral composed of mica and feldspar.

B. A meteorite is material from outer space, that has struck the earth's surface.

C. A meteorite is an element that has properties of both metals and nonmetals.

D. A meteorite is a very small unit of length measurement.

113. A white flower is crossed with a red flower. Which of the following is a sign of incomplete dominance?

A. Pink flowers.

B. Red flowers.

C. White flowers.

D. No flowers.

114. What is the source for most of the United States' drinking water?

A. Desalinated ocean water.

B. Surface water (lakes, streams, mountain runoff).

C. Rainfall into municipal reservoirs.

D. Groundwater.

115. Which is the correct sequence of insect development?

A. Egg, pupa, larva, adult.

B. Egg, larva, pupa, adult.

C. Egg, adult, larva, pupa.

D. Pupa, egg, larva, adult.

116. A wrasse (fish) cleans the teeth of other fish by eating away plaque. This is an example of ___________ between the fish.

A. parasitism.

B. symbiosis (mutualism).

C. competition.

D. predation.

117. What is the main obstacle to using nuclear fusion for obtaining electricity?

A. Nuclear fusion produces much more pollution than nuclear fission.

B. There is no obstacle; most power plants us nuclear fusion today.

C. Nuclear fusion requires very high temperature and activation energy.

D. The fuel for nuclear fusion is extremely expensive.

118. Which of the following is a true statement about radiation exposure and air travel?

A. Air travel exposes humans to radiation, but the level is not significant for most people.

B. Air travel exposes humans to so much radiation that it is recommended as a method of cancer treatment.

C. Air travel does not expose humans to radiation.

D. Air travel may or may not expose humans to radiation, but it has not yet been determined.

119. Which process(es) result(s) in a haploid chromosome number?

A. Mitosis.

B. Meiosis.

C. Both mitosis and meiosis.

D. Neither mitosis nor meiosis.

120. Which of the following is NOT a member of Kingdom Fungi?

A. Mold.

B. Blue-green algae.

C. Mildew.

D. Mushrooms.

121. Which of the following organisms use spores to reproduce?

A. Fish.

B. Flowering plants.

C. Conifers.

D. Ferns.

122. What is the main difference between the 'condensation hypothesis' and the 'tidal hypothesis' for the origin of the solar system?

A. The tidal hypothesis can be tested, but the condensation hypothesis cannot.

B. The tidal hypothesis proposes a near collision of two stars pulling on each other, but the condensation hypothesis proposes condensation of rotating clouds of dust and gas.

C. The tidal hypothesis explains how tides began on planets such as Earth, but the condensation hypothesis explains how water vapor became liquid on Earth.

D. The tidal hypothesis is based on Aristotelian physics, but the condensation hypothesis is based on Newtonian mechanics.

123. Which of the following units is NOT a measure of distance?

A. AU (astronomical unit).

B. Light year.

C. Parsec.

D. Lunar year.

124. The salinity of ocean water is closest to ____________ .

A. 0.035 %

B. 0.35 %

C. 3.5 %

D. 35 %

125. Which of the following will not change in a chemical reaction?

A. Number of moles of products.

B. Atomic number of one of the reactants.

C. Mass (in grams) of one of the reactants.

D. Rate of reaction.

Answer Key

1.	B	26.	A	51.	D	76.	A	101.	C
2.	A	27.	B	52.	D	77.	B	102.	B
3.	A	28.	D	53.	B	78.	B	103.	C
4.	C	29.	B	54.	B	79.	A	104.	D
5.	B	30.	B	55.	A	80.	A	105.	D
6.	C	31.	C	56.	B	81.	B	106.	B
7.	D	32.	C	57.	B	82.	A	107.	D
8.	C	33.	D	58.	A	83.	D	108.	B
9.	C	34.	A	59.	A	84.	B	109.	C
10.	B	35.	B	60.	C	85.	C	110.	B
11.	A	36.	D	61.	C	86.	B	111.	A
12.	C	37.	A	62.	A	87.	C	112.	B
13.	C	38.	B	63.	B	88.	C	113.	A
14.	B	39.	B	64.	A	89.	C	114.	D
15.	B	40.	A	65.	C	90.	C	115.	B
16.	B	41.	D	66.	D	91.	C	116.	B
17.	A	42.	A	67.	C	92.	C	117.	C
18.	C	43.	B	68.	D	93.	A	118.	A
19.	C	44.	A	69.	B	94.	A	119.	B
20.	D	45.	C	70.	C	95.	B	120.	B
21.	C	46.	C	71.	B	96.	D	121.	D
22.	B	47.	B	72.	B	97.	B	122.	B
23.	A	48.	D	73.	A	98.	D	123.	D
24.	A	49.	B	74.	D	99.	C	124.	C
25.	A	50.	A	75.	D	100.	A	125.	B

Rationales with Sample Questions

1. After an experiment, the scientist states that s/he believes a change in color is due to a change in pH. This is an example of

A. observing.

B. inferring.

C. measuring.

D. classifying.

B. Inferring.

To answer this question, note that the scientist has observed a change in color, and has then made a guess as to its reason. This is an example of inferring. The scientist has not measured or classified in this case. Although s/he has observed [the color change], the explanation of this observation is **inferring (B).**

2. When is a hypothesis formed?

A. Before the data is taken.

B. After the data is taken.

C. After the data is analyzed.

D. While the data is being graphed.

A. Before the data is taken.

A hypothesis is an educated guess, made before undertaking an experiment. The hypothesis is then evaluated based on the observed data. Therefore, the hypothesis must be formed before the data is taken, not during or after the experiment. This is consistent only with **answer (A).**

3. Who determines the laws regarding the use of safety glasses in the classroom?

A. The state government.

B. The school site.

C. The federal government.

D. The local district.

A. The state government.

Health and safety regulations are set by the state government, and apply to all school districts. Federal regulations may accompany specific federal grants, and local districts or school sites may enact local guidelines that are stricter than the state standards. All schools, however, must abide by safety precautions as set by state government. This is consistent only with **answer (A).**

4. If one inch equals 2.54 centimeters, how many millimeters are in 1.5 feet? (Approximately)

A. 18

B. 1800

C. 460

D. 4600

C. 460

To solve this problem, note that if one inch is 2.54 centimeters, then 1.5 feet (which is 18 inches), must be (18)(2.54) centimeters, i.e. approximately 46 centimeters. Because there are ten millimeters in a centimeter, this is approximately 460 millimeters:

$(1.5 \text{ ft})\ (12 \text{ in/ft})\ (2.54 \text{ cm/in})\ (10 \text{ mm/cm}) = (1.5)\ (12)\ (2.54)\ (10) \text{ mm} = 457.2 \text{ mm}$

This is consistent only with **answer (C).**

5. Which of the following instruments measures wind speed?

A. Barometer.

B. Anemometer.

C. Thermometer.

D. Weather Vane.

B. Anemometer.

An anemometer is a device to measure wind speed, while a barometer measures pressure, a thermometer measures temperature, and a weather vane indicates wind direction. This is consistent only with **answer (B).**

If you chose "barometer," here is an old physics joke to console you:

A physics teacher asks a student the following question:
"Suppose you want to find out the height of a building, and the only tool you have is a barometer. How could you find out the height?"
(The teacher hopes that the student will remember that pressure is inversely proportional to height, and will measure the pressure at the top of the building and then use the data to calculate the height of the building.)
"Well," says the student, "I could tie a string to the barometer and lower it from the top of the building, and then measure the amount of string required."
"You could," answers the teacher, "but try to think of a method that uses your physics knowledge from our class."
"All right," replies the student, "I could drop the barometer from the roof and measure the time it takes to fall, and then use free-fall equations to calculate the height from which it fell."
"Yes," says the teacher, "but what about using the barometer per se?"
"Oh," answers the student, "I could find the building superintendent, and offer to exchange the barometer for a set of blueprints, and look up the height!"

6. Sonar works by

A. timing how long it takes sound to reach a certain speed.

B. bouncing sound waves between two metal plates.

C. bouncing sound waves off an object and timing how long it takes for the sound to return.

D. evaluating the motion and amplitude of sound.

C. Bouncing sound waves off an object and timing how long it takes for the sound to return.

Sonar is used to measure distances. Sound waves are sent out, and the time is measured for the sound to hit an obstacle and bounce back. By using the known speed of sound, observers (or machines) can calculate the distance to the obstacle. This is consistent only with **answer (C).**

7. The measure of the pull of Earth's gravity on an object is called

A. mass number.

B. atomic number.

C. mass.

D. weight.

D. Weight.

To answer this question, recall that mass number is the total number of protons and neutrons in an atom, atomic number is the number of protons in an atom, and mass is the amount of matter in an object. The only remaining **choice is (D)**, weight, which is correct because weight is the force of gravity on an object.

8. Which reaction below is a decomposition reaction?

A. $HCl + NaOH \rightarrow NaCl + H_2O$

B. $C + O_2 \rightarrow CO_2$

C. $2H_2O \rightarrow 2H_2 + O_2$

D. $CuSO_4 + Fe \rightarrow FeSO_4 + Cu$

C. $2H_2O \rightarrow 2H_2 + O_2$

To answer this question, recall that a decomposition reaction is one in which there are fewer reactants (on the left) than products (on the right). This is consistent only with **answer (C).** Meanwhile, note that answer (A) shows a double-replacement reaction (in which two sets of ions switch bonds), answer (B) shows a synthesis reaction (in which there are fewer products than reactants), and answer (D) shows a single-replacement reaction (in which one substance replaces another in its bond, but the other does not get a new bond).

9. The Law of Conservation of Energy states that

A. there must be the same number of products and reactants in any chemical equation.

B. objects always fall toward large masses such as planets.

C. energy is neither created nor destroyed, but may change form.

D. lights must be turned off when not in use, by state regulation.

C. **Energy is neither created nor destroyed, but may change form.**

Answer (C) is a summary of the Law of Conservation of Energy (for non-nuclear reactions). In other words, energy can be transformed into various forms such as kinetic, potential, electric, or heat energy, but the total amount of energy remains constant. Answer (A) is untrue, as demonstrated by many synthesis and decomposition reactions. Answers (B) and (D) may be sensible, but they are not relevant in this case. Therefore, the **answer is (C).**

10. Which parts of an atom are located inside the nucleus?

A. Protons and Electrons.

B. Protons and Neutrons.

C. Protons only.

D. Neutrons only.

B. Protons and Neutrons.

Protons and neutrons are located in the nucleus, while electrons move around outside the nucleus. This is consistent only with **answer (B)**.

11. The elements in the modern Periodic Table are arranged

A. in numerical order by atomic number.

B. randomly.

C. in alphabetical order by chemical symbol.

D. in numerical order by atomic mass.

A. In numerical order by atomic number.

Although the first periodic tables were arranged by atomic mass, the modern table is arranged by atomic number, i.e. the number of protons in each element. (This allows the element list to be complete and unique.) The elements are not arranged either randomly or in alphabetical order. The answer to this question is **therefore (A)**.

12. Carbon bonds with hydrogen by

A. ionic bonding.

B. non-polar covalent bonding.

C. polar covalent bonding.

D. strong nuclear force.

C. Polar covalent bonding.

Each carbon atom contains four valence electrons, while each hydrogen atom contains one valence electron. A carbon atom can bond with one or more hydrogen atoms, such that two electrons are shared in each bond. This is covalent bonding, because the electrons are shared. (In ionic bonding, atoms must gain or lose electrons to form ions. The ions are then electrically attracted in oppositely-charged pairs.) Covalent bonds are always polar when between two non-identical atoms, so this bond must be polar. ("Polar" means that the electrons are shared unequally, forming a pair of partial charges, i.e. poles.) In any case, the strong nuclear force is not relevant to this problem. The answer to this question is **therefore (C).**

13. Vinegar is an example of a

A. strong acid.

B. strong base.

C. weak acid.

D. weak base.

C. Weak acid.

The main ingredient in vinegar is acetic acid, a weak acid. Vinegar is a useful acid in science classes, because it makes a frothy reaction with bases such as baking soda (e.g. in the quintessential volcano model). Vinegar is not a strong acid, such as hydrochloric acid, because it does not dissociate as fully or cause as much corrosion. It is not a base. Therefore, the **answer is (C)**.

14. Which of the following is not a nucleotide?

A. Adenine.

B. Alanine.

C. Cytosine.

D. Guanine.

B. **Alanine.**

Alanine is an amino acid. Adenine, cytosine, guanine, thymine, and uracil are nucleotides. The correct **answer is (B).**

15. When measuring the volume of water in a graduated cylinder, where does one read the measurement?

A. At the highest point of the liquid.

B. At the bottom of the meniscus curve.

C. At the closest mark to the top of the liquid.

D. At the top of the plastic safety ring.

B. **At the bottom of the meniscus curve.**

To measure water in glass, you must look at the top surface at eye-level, and ascertain the location of the bottom of the meniscus (the curved surface at the top of the water). The meniscus forms because water molecules adhere to the sides of the glass, which is a slightly stronger force than their cohesion to each other. This leads to a U-shaped top of the liquid column, the bottom of which gives the most accurate volume measurement. (Other liquids have different forces, e.g. mercury in glass, which has a convex meniscus.) This is consistent only with **answer (B).**

16. A duck's webbed feet are examples of

A. mimicry.

B. structural adaptation.

C. protective resemblance.

D. protective coloration.

B. Structural adaptation.

Ducks (and other aquatic birds) have webbed feet, which makes them more efficient swimmers. This is most likely due to evolutionary patterns where webbed-footed-birds were more successful at feeding and reproducing, and eventually became the majority of aquatic birds. Because the structure of the duck adapted to its environment over generations, this is termed 'structural adaptation'. Mimicry, protective resemblance, and protective coloration refer to other evolutionary mechanisms for survival. The answer to this question is **therefore (B)**.

17. What cell organelle contains the cell's stored food?

A. Vacuoles.

B. Golgi Apparatus.

C. Ribosomes.

D. Lysosomes.

A. Vacuoles.

In a cell, the sub-parts are called organelles. Of these, the vacuoles hold stored food (and water and pigments). The Golgi Apparatus sorts molecules from other parts of the cell; the ribosomes are sites of protein synthesis; the lysosomes contain digestive enzymes. This is consistent only with **answer (A).**

18. The first stage of mitosis is called

A. telophase.

B. anaphase.

C. prophase.

D. mitophase.

C. Prophase.

In mitosis, the division of somatic cells, prophase is the stage where the cell enters mitosis. The four stages of mitosis, in order, are: prophase, metaphase, anaphase, and telophase. ("Mitophase" is not one of the steps.) During prophase, the cell begins the nonstop process of division. Its chromatin condenses, its nucleolus disappears, the nuclear membrane breaks apart, mitotic spindles form, its cytoskeleton breaks down, and centrioles push the spindles apart. Note that interphase, the stage where chromatin is loose, chromosomes are replicated, and cell metabolism is occurring, is technically not a stage of mitosis; it is a precursor to cell division.

19. The Doppler Effect is associated most closely with which property of waves?

A. Amplitude.

B. Wavelength.

C. Frequency.

D. Intensity.

C. Frequency.

The Doppler Effect accounts for an apparent increase in frequency when a wave source moves toward a wave receiver or apparent decrease in frequency when a wave source moves away from a wave receiver. (Note that the receiver could also be moving toward or away from the source.) As the wave fronts are released, motion toward the receiver mimics more frequent wave fronts, while motion away from the receiver mimics less frequent wave fronts. Meanwhile, the amplitude, wavelength, and intensity of the wave are not as relevant to this process (although moving closer to a wave source makes it seem more intense). The **answer to this question is therefore (C)**.

20. Viruses are responsible for many human diseases including all of the following EXCEPT

A. influenza.

B. A.I.D.S.

C. the common cold.

D. strep throat.

D. Strep throat.

Influenza, A.I.D.S., and the “common cold” (rhinovirus infection), are all caused by viruses. (This is the reason that doctors should not be pressured to prescribe antibiotics for colds or ‘flu—i.e. they will not be effective since the infections are not bacterial.) Strep throat (properly called ‘streptococcal throat’ and caused by streptococcus bacteria) is not a virus, but a bacterial infection. Thus, the **answer is (D)**.

21. A series of experiments on pea plants formed by _________ showed that two invisible markers existed for each trait, and one marker dominated the other.

A. Pasteur.

B. Watson and Crick.

C. Mendel.

D. Mendeleev.

C. Mendel.

Gregor Mendel was a ninteenth-century Austrian botanist, who derived “laws” governing inherited traits. His work led to the understanding of dominant and recessive traits, carried by biological markers. Mendel cross-bred different kinds of pea plants with varying features and observed the resulting new plants. He showed that genetic characteristics are not passed identically from one generation to the next. (Pasteur, Watson, Crick, and Mendeleev were other scientists with different specialties.) This is consistent only with **answer (C)**.

22. Formaldehyde should not be used in school laboratories for the following reason:

A. it smells unpleasant.

B. it is a known carcinogen.

C. it is expensive to obtain.

D. it is an explosive.

B. It is a known carcinogen.

Formaldehyde is a known carcinogen, so it is too dangerous for use in schools. In general, teachers should not use carcinogens in school laboratories. Although formaldehyde also smells unpleasant, a smell alone is not a definitive marker of danger. For example, many people find the smell of vinegar to be unpleasant, but vinegar is considered a very safe classroom/laboratory chemical. Furthermore, some odorless materials are toxic. Formaldehyde is neither particularly expensive nor explosive. Thus, the **answer is (B)**.

23. Amino acids are carried to the ribosome in protein synthesis by:

A. transfer RNA (tRNA).

B. messenger RNA (mRNA).

C. ribosomal RNA (rRNA).

D. transformation RNA (trRNA).

A. Transfer RNA (tRNA).

The job of tRNA is to carry and position amino acids to/on the ribosomes. mRNA copies DNA code and brings it to the ribosomes; rRNA is in the ribosome itself. There is no such thing as trRNA. Thus, the **answer is (A)**.

24. When designing a scientific experiment, a student considers all the factors that may influence the results. The process goal is to

A. recognize and manipulate independent variables.

B. recognize and record independent variables.

C. recognize and manipulate dependent variables.

D. recognize and record dependent variables.

A. Recognize and manipulate independent variables.

When a student designs a scientific experiment, s/he must decide what to measure, and what independent variables will play a role in the experiment. S/he must determine how to manipulate these independent variables to refine his/her procedure and to prepare for meaningful observations. Although s/he will eventually record dependent variables (D), this does not take place during the experimental design phase. Although the student will likely recognize and record the independent variables (B), this is not the process goal, but a helpful step in manipulating the variables. It is unlikely that the student will manipulate dependent variables directly in his/her experiment (C), or the data would be suspect. Thus, the **answer is (A)**.

25. Since ancient times, people have been entranced with bird flight. What is the key to bird flight?

A. Bird wings are a particular shape and composition.

B. Birds flap their wings quickly enough to propel themselves.

C. Birds take advantage of tailwinds.

D. Birds take advantage of crosswinds.

A. Bird wings are a particular shape and composition.

Bird wings are shaped for wide area, and their bones are very light. This creates a large surface-area-to-mass ratio, enabling birds to glide in air. Birds do flap their wings and float on winds, but none of these is the main reason for their flight ability. Thus, the **answer is (A)**.

26. Laboratory researchers have classified fungi as distinct from plants because the cell walls of fungi

A. contain chitin.

B. contain yeast.

C. are more solid.

D. are less solid.

A. Contain chitin.

Kingdom Fungi consists of organisms that are eukaryotic, multicellular, absorptive consumers. They have a chitin cell wall, which is the only universally present feature in fungi that is never present in plants. Thus, the **answer is (A)**.

27. In a fission reactor, "heavy water" is used to

A. terminate fission reactions.

B. slow down neutrons and moderate reactions.

C. rehydrate the chemicals.

D. initiate a chain reaction.

B. Slow down neutrons and moderate reactions.

"Heavy water" is used in a nuclear [fission] reactor to slow down neutrons, controlling and moderating the nuclear reactions. It does not terminate the reaction, and it does not initiate the reaction. Also, although the reactor takes advantage of water's other properties (e.g. high specific heat for cooling), the water does not "rehydrate" the chemicals. Therefore, the **answer is (B)**.

28. The transfer of heat by electromagnetic waves is called

A. conduction.

B. convection.

C. phase change.

D. radiation.

D. Radiation.

Heat transfer via electromagnetic waves (which can occur even in a vacuum) is called radiation. (Heat can also be transferred by direct contact (conduction), by fluid current (convection), and by matter changing phase, but these are not relevant here.) The answer to this question is **therefore (D)**.

29. When heat is added to most solids, they expand. Why is this the case?

A. The molecules get bigger.

B. The faster molecular motion leads to greater distance between the molecules.

C. The molecules develop greater repelling electric forces.

D. The molecules form a more rigid structure.

B. The faster molecular motion leads to greater distance between the molecules.

The atomic theory of matter states that matter is made up of tiny, rapidly moving particles. These particles move more quickly when warmer, because temperature is a measure of average kinetic energy of the particles. Warmer molecules therefore move further away from each other, with enough energy to separate from each other more often and for greater distances. The individual molecules do not get bigger, by conservation of mass, eliminating answer (A). The molecules do not develop greater repelling electric forces, eliminating answer (C). Occasionally, molecules form a more rigid structure when becoming colder and freezing (such as water)—but this gives rise to the exceptions to heat expansion, so it is not relevant here, eliminating answer (D). Therefore, the **answer is (B)**.

30. The force of gravity on earth causes all bodies in free fall to

A. fall at the same speed.

B. accelerate at the same rate.

C. reach the same terminal velocity.

D. move in the same direction.

B. Accelerate at the same rate.

Gravity causes approximately the same acceleration on all falling bodies close to earth's surface. (It is only "approximately" because there are very small variations in the strength of earth's gravitational field.) More massive bodies continue to accelerate at this rate for longer, before their air resistance is great enough to cause terminal velocity, so answers (A) and (C) are eliminated. Bodies on different parts of the planet move in different directions (always toward the center of mass of earth), so answer (D) is eliminated. Thus, the **answer is (B)**.

31. Sound waves are produced by

A. pitch.

B. noise.

C. vibrations.

D. sonar.

C. Vibrations.

Sound waves are produced by a vibrating body. The vibrating object moves forward and compresses the air in front of it, then reverses direction so that the pressure on the air is lessened and expansion of the air molecules occurs. The vibrating air molecules move back and forth parallel to the direction of motion of the wave as they pass the energy from adjacent air molecules closer to the source to air molecules farther away from the source. Therefore, the **answer is (C)**.

32. Resistance is measured in units called

A. watts.

B. volts.

C. ohms.

D. current.

C. Ohms.

A watt is a unit of energy. Potential difference is measured in a unit called the volt. Current is the number of electrons per second that flow past a point in a circuit. An ohm is the unit for resistance. The correct **answer is (C).**

33. Sound can be transmitted in all of the following EXCEPT

A. air.

B. water.

C. diamond.

D. a vacuum.

D. A vacuum.

Sound, a longitudinal wave, is transmitted by vibrations of molecules. Therefore, it can be transmitted through any gas, liquid, or solid. However, it cannot be transmitted through a vacuum, because there are no particles present to vibrate and bump into their adjacent particles to transmit the waves. This is consistent only with **answer (D)**. (It is interesting also to note that sound is actually faster in solids and liquids than in air.)

34. As a train approaches, the whistle sounds

A. higher, because it has a higher apparent frequency.

B. lower, because it has a lower apparent frequency.

C. higher, because it has a lower apparent frequency.

D. lower, because it has a higher apparent frequency.

A. Higher, because it has a higher apparent frequency.

By the Doppler effect, when a source of sound is moving toward an observer, the wave fronts are released closer together, i.e. with a greater apparent frequency. Higher frequency sounds are higher in pitch. This is consistent only with **answer (A)**.

35. The speed of light is different in different materials. This is responsible for

A. interference.

B. refraction.

C. reflection.

D. relativity.

B. Refraction.

Refraction (B) is the bending of light because it hits a material at an angle wherein it has a different speed. (This is analogous to a cart rolling on a smooth road. If it hits a rough patch at an angle, the wheel on the rough patch slows down first, leading to a change in direction.) Interference (A) is when light waves interfere with each other to form brighter or dimmer patterns; reflection (C) is when light bounces off a surface; relativity (D) is a general topic related to light speed and its implications, but not specifically indicated here. Therefore, the **answer is (B)**.

36. A converging lens produces a real image _______________

A. always.

B. never.

C. when the object is within one focal length of the lens.

D. when the object is further than one focal length from the lens.

D. When the object is further than one focal length from the lens.

A converging lens produces a real image whenever the object is far enough from the lens (outside one focal length) so that the rays of light from the object can hit the lens and be focused into a real image on the other side of the lens. When the object is closer than one focal length from the lens, rays of light do not converge on the other side; they diverge. This means that only a virtual image can be formed, i.e. the theoretical place where those diverging rays would have converged if they had originated behind the object. Thus, the correct **answer is (D)**.

37. The electromagnetic radiation with the longest wave length is/are ____________

A. radio waves.

B. red light.

C. X-rays.

D. ultraviolet light.

A. Radio waves.

As one can see on a diagram of the electromagnetic spectrum, radio waves have longer wave lengths (and smaller frequencies) than visible light, which in turn has longer wave lengths than ultraviolet or X-ray radiation. If you did not remember this sequence, you might recall that wave length is inversely proportional to frequency, and that radio waves are considered much less harmful (less energetic, i.e. lower frequency) than ultraviolet or X-ray radiation. The correct answer is **therefore (A)**.

38. Under a 440 power microscope, an object with diameter 0.1 millimeter appears to have diameter ______________

A. 4.4 millimeters.

B. 44 millimeters.

C. 440 millimeters.

D. 4400 millimeters.

B. 44 millimeters.

To answer this question, recall that to calculate a new length, you multiply the original length by the magnification power of the instrument. Therefore, the 0.1 millimeter diameter is multiplied by 440. This equals 44, so the image appears to be 44 millimeters in diameter. You could also reason that since a 440 power microscope is considered a "high power" microscope, you would expect a 0.1 millimeter object to appear a few centimeters long. Therefore, the correct **answer is (B)**.

39. To separate blood into blood cells and plasma involves the process of

A. electrophoresis.

B. centrifugation.

C. spectrophotometry.

D. chromatography.

B. Centrifugation.

Electrophoresis uses electrical charges of molecules to separate them according to their size. Spectrophotometry uses percent light absorbance to measure a color change, thus giving qualitative data a quantitative value. Chromatography uses the principles of capillarity to separate substances. Centrifugation involves spinning substances at a high speed. The more dense part of a solution will settle to the bottom of the test tube, where the lighter material will stay on top. The **answer is (B).**

40. Experiments may be done with any of the following animals except

A. birds.

B. invertebrates.

C .lower order life.

D. frogs.

A. Birds.

No dissections may be performed on living mammalian vertebrates or birds. Lower order life and invertebrates may be used. Biological experiments may be done with all animals except mammalian vertebrates or birds. Therefore the **answer is (A).**

41. For her first project of the year, a student is designing a science experiment to test the effects of light and water on plant growth. You should recommend that she ________________

A. manipulate the temperature also.

B. manipulate the water pH also.

C. determine the relationship between light and water unrelated to plant growth.

D. omit either water or light as a variable.

D. **Omit either water or light as a variable.**

As a science teacher for middle-school-aged kids, it is important to reinforce the idea of 'constant' vs. 'variable' in science experiments. At this level, it is wisest to have only one variable examined in each science experiment. (Later, students can hold different variables constant while investigating others.) Therefore it is counterproductive to add in other variables (answers (A)or (B)). It is also irrelevant to determine the light-water interactions aside from plant growth (C). So the only possible **answer is (D)**.

42. In a laboratory report, what is the abstract?

A. The abstract is a summary of the report, and is the first section of the report.

B. The abstract is a summary of the report, and is the last section of the report.

C. The abstract is predictions for future experiments, and is the first section of the report.

D. The abstract is predictions for future experiments, and is the last section of the report.

A. The abstract is a summary of the report, and is the first section of the report.

In a laboratory report, the abstract is the section that summarizes the entire report (often containing one representative sentence from each section). It appears at the very beginning of the report, even before the introduction, often on its own page (instead of a title page). This format is consistent with articles in scientific journals. Therefore, the **answer is (A).**

43. What is the scientific method?

A. It is the process of doing an experiment and writing a laboratory report.

B. It is the process of using open inquiry and repeatable results to establish theories.

C. It is the process of reinforcing scientific principles by confirming results.

D. It is the process of recording data and observations.

B. It is the process of using open inquiry and repeatable results to establish theories.

Scientific research often includes elements from answers (A), (C), and (D), but the basic underlying principle of the scientific method is that people ask questions and do repeatable experiments to answer those questions and develop informed theories of why and how things happen. Therefore, the best **answer is (B).**

44. Identify the control in the following experiment: A student had four corn plants and was measuring photosynthetic rate (by measuring growth mass). Half of the plants were exposed to full (constant) sunlight, and the other half were kept in 50% (constant) sunlight.

A. The control is a set of plants grown in full (constant) sunlight.

B. The control is a set of plants grown in 50% (constant) sunlight.

C. The control is a set of plants grown in the dark.

D. The control is a set of plants grown in a mixture of natural levels of sunlight.

A. The control is a set of plants grown in full (constant) sunlight.

In this experiment, the goal was to measure how two different amounts of sunlight affected plant growth. The control in any experiment is the 'base case,' or the usual situation without a change in variable. Because the control must be studied alongside the variable, answers (C) and (D) are omitted (because they were not in the experiment). The **better answer of (A) and (B) is (A)**, because usually plants are assumed to have the best growth and their usual growing circumstances in full sunlight. This is particularly true for crops like the corn plants in this question.

45. In an experiment measuring the growth of bacteria at different temperatures, what is the independent variable?

A. Number of bacteria.

B. Growth rate of bacteria.

C. Temperature.

D. Light intensity.

C. Temperature.

To answer this question, recall that the independent variable in an experiment is the entity that is changed by the scientist, in order to observe the effects (the dependent variable(s)). In this experiment, temperature is changed in order to measure growth of bacteria, so **(C) is the answer**. Note that answer (A) is the dependent variable, and neither (B) nor (D) is directly relevant to the question.

46. A scientific law ____________

A. proves scientific accuracy.

B. may never be broken.

C. may be revised in light of new data.

D. is the result of one excellent experiment.

C. May be revised in light of new data.

A scientific law is the same as a scientific theory, except that it has lasted for longer, and has been supported by more extensive data. Therefore, such a law may be revised in light of new data, and may be broken by that new data. Furthermore, a scientific law is always the result of many experiments, and never 'proves' anything but rather is implied or supported by various results. Therefore, the **answer must be (C).**

47. Which is the correct order of methodology?

1. **collecting data**
2. **planning a controlled experiment**
3. **drawing a conclusion**
4. **hypothesizing a result**
5. **re-visiting a hypothesis to answer a question**

A. 1,2,3,4,5

B. 4,2,1,3,5

C. 4,5,1,3,2

D. 1,3,4,5,2

B. 4,2,1,3,5

The correct methodology for the scientific method is first to make a meaningful hypothesis (educated guess), then plan and execute a controlled experiment to test that hypothesis. Using the data collected in that experiment, the scientist then draws conclusions and attempts to answer the original question related to the hypothesis. This is consistent only with **answer (B)**.

48. Which is the most desirable tool to use to heat substances in a middle school laboratory?

A. Alcohol burner.

B. Freestanding gas burner.

C. Bunsen burner.

D. Hot plate.

D. Hot plate.

Due to safety considerations, the use of open flame should be minimized, so a hot plate is the best choice. Any kind of burner may be used with proper precautions, but it is difficult to maintain a completely safe middle school environment. Therefore, the best **answer is (D)**.

49. Newton's Laws are taught in science classes because __________________

A. they are the correct analysis of inertia, gravity, and forces.

B. they are a close approximation to correct physics, for usual Earth conditions.

C. they accurately incorporate Relativity into studies of forces.

D. Newton was a well-respected scientist in his time.

B. They are a close approximation to correct physics, for usual Earth conditions.

Although Newton's Laws are often taught as fully correct for inertia, gravity, and forces, it is important to realize that Einstein's work (and that of others) has indicated that Newton's Laws are reliable only at speeds much lower than that of light. This is reasonable, though, for most middle- and high-school applications. At speeds close to the speed of light, Relativity considerations must be used. Therefore, the only correct **answer is (B)**.

50. Which of the following is most accurate?

A. Mass is always constant; Weight may vary by location.

B. Mass and Weight are both always constant.

C. Weight is always constant; Mass may vary by location.

D. Mass and Weight may both vary by location.

A. **Mass is always constant; Weight may vary by location**.

When considering situations exclusive of nuclear reactions, mass is constant (mass, the amount of matter in a system, is conserved). Weight, on the other hand, is the force of gravity on an object, which is subject to change due to changes in the gravitational field and/or the location of the object. Thus, the **best answer is (A)**.

51. Chemicals should be stored __________________

A. in the principal's office.

B. in a dark room.

C. in an off-site research facility.

D. according to their reactivity with other substances.

D. **According to their reactivity with other substances.**

Chemicals should be stored with other chemicals of similar properties (e.g. acids with other acids), to reduce the potential for either hazardous reactions in the storeroom, or mistakes in reagent use. Certainly, chemicals should not be stored in anyone's office, and the light intensity of the room is not very important because light-sensitive chemicals are usually stored in dark containers. In fact, good lighting is desirable in a storeroom, so that labels can be read easily. Chemicals may be stored off-site, but that makes their use inconvenient. Therefore, the best **answer is (D)**.

52. Which of the following is the worst choice for a school laboratory activity?

A. A genetics experiment tracking the fur color of mice.

B. Dissection of a preserved fetal pig.

C. Measurement of goldfish respiration rate at different temperatures.

D. Pithing a frog to watch the circulatory system.

D. Pithing a frog to watch the circulatory system.

While any use of animals (alive or dead) must be done with care to respect ethics and laws, it is possible to perform choices (A), (B), or (C) with due care. (Note that students will need significant assistance and maturity to perform these experiments.) However, modern practice precludes pithing animals (causing partial brain death while allowing some systems to function), as inhumane. Therefore, the answer to this **question is (D)**.

53. Who should be notified in the case of a serious chemical spill?

A. The custodian.

B. The fire department or other municipal authority.

C. The science department chair.

D. The School Board.

B. The fire department or other municipal authority.

Although the custodian may help to clean up laboratory messes, and the science department chair should be involved in discussions of ways to avoid spills, a serious chemical spill may require action by the fire department or other trained emergency personnel. It is best to be safe by notifying them in case of a serious chemical accident. Therefore, the **best answer is (B)**.

54. A scientist exposes mice to cigarette smoke, and notes that their lungs develop tumors. Mice that were not exposed to the smoke do not develop as many tumors. Which of the following conclusions may be drawn from these results?

I. Cigarette smoke causes lung tumors.
II. Cigarette smoke exposure has a positive correlation with lung tumors in mice.
III. Some mice are predisposed to develop lung tumors.
IV. Mice are often a good model for humans in scientific research.

A. I and II only.

B. II only.

C. I , II, and III only.

D. II and IV only.

B. II only.

Although cigarette smoke has been found to cause lung tumors (and many other problems), this particular experiment shows only that there is a positive correlation between smoke exposure and tumor development in these mice. It may be true that some mice are more likely to develop tumors than others, which is why a control group of identical mice should have been used for comparison. Mice are often used to model human reactions, but this is as much due to their low financial and emotional cost as it is due to their being a "good model" for humans. Therefore, the **answer must be (B)**.

55. In which situation would a science teacher be legally liable?

A. The teacher leaves the classroom for a telephone call and a student slips and injures him/herself.

B. A student removes his/her goggles and gets acid in his/her eye.

C. A faulty gas line in the classroom causes a fire.

D. A student cuts him/herself with a dissection scalpel.

A. The teacher leaves the classroom for a telephone call and a student slips and injures him/herself.

Teachers are required to exercise a "reasonable duty of care" for their students. Accidents may happen (e.g. (D)), or students may make poor decisions (e.g. (B)), or facilities may break down (e.g. (C)). However, the teacher has the responsibility to be present and to do his/her best to create a safe and effective learning environment. Therefore, the **answer is (A)**.

56. Which of these is the best example of 'negligence'?

A. A teacher fails to give oral instructions to those with reading disabilities.

B. A teacher fails to exercise ordinary care to ensure safety in the classroom.

C. A teacher displays inability to supervise a large group of students.

D. A teacher reasonably anticipates that an event may occur, and plans accordingly.

B. A teacher fails to exercise ordinary care to ensure safety in the classroom.

'Negligence' is the failure to "exercise ordinary care" to ensure an appropriate and safe classroom environment. It is best for a teacher to meet all special requirements for disabled students, and to be good at supervising large groups. However, if a teacher can prove that s/he has done a reasonable job to ensure a safe and effective learning environment, then it is unlikely that she/he would be found negligent. Therefore, **the answer is (B)**.

57. Which item should always be used when handling glassware?

A. Tongs.

B. Safety goggles.

C. Gloves.

D. Buret stand.

B. Safety goggles.

Safety goggles are the single most important piece of safety equipment in the laboratory, and should be used any time a scientist is using glassware, heat, or chemicals. Other equipment (e.g. tongs, gloves, or even a buret stand) has its place for various applications. However, the most important is safety goggles. Therefore, the **answer is (B)**.

58. Which of the following is NOT a necessary characteristic of living things?

A. Movement.

B. Reduction of local entropy.

C. Ability to cause local energy form changes.

D. Reproduction.

A. Movement.

There are many definitions of "life," but in all cases, a living organism reduces local entropy, changes chemical energy into other forms, and reproduces. Not all living things move, however, so the correct **answer is (A)**.

59. What are the most significant and prevalent elements in the biosphere?

A. Carbon, Hydrogen, Oxygen, Nitrogen, Phosphorus.

B. Carbon, Hydrogen, Sodium, Iron, Calcium.

C. Carbon, Oxygen, Sulfur, Manganese, Iron.

D. Carbon, Hydrogen, Oxygen, Nickel, Sodium, Nitrogen.

A. Carbon, Hydrogen, Oxygen, Nitrogen, Phosphorus.

Organic matter (and life as we know it) is based on carbon atoms, bonded to hydrogen and oxygen. Nitrogen and phosphorus are the next most significant elements, followed by sulfur and then trace nutrients such as iron, sodium, calcium, and others. Therefore, the **answer is (A)**. If you know that the formula for any carbohydrate contains carbon, hydrogen, and oxygen, that will help you narrow the choices to (A) and (D) in any case.

60. All of the following measure energy EXCEPT for __________

A. joules.

B. calories.

C. watts.

D. ergs.

C. Watts.

Energy units must be dimensionally equivalent to (force)x(length), which equals (mass)x(length squared)/(time squared). Joules, Calories, and Ergs are all metric measures of energy. Joules are the SI units of energy, while Calories are used to allow water to have a Specific Heat of one unit. Ergs are used in the 'cgs' (centimeter-gram-second) system, for smaller quantities. Watts, however, are units of power, i.e. Joules per Second. Therefore, the **answer is (C)**.

61. Identify the correct sequence of organization of living things from lower to higher order:

A. Cell, Organelle, Organ, Tissue, System, Organism.

B. Cell, Tissue, Organ, Organelle, System, Organism.

C. Organelle, Cell, Tissue, Organ, System, Organism.

D. Organelle, Tissue, Cell, Organ, System, Organism.

C. Organelle, Cell, Tissue, Organ, System, Organism.

Organelles are parts of the cell; cells make up tissue, which makes up organs. Organs work together in systems (e.g. the respiratory system), and the organism is the living thing as a whole. Therefore, the **answer must be (C)**.

62. Which kingdom is comprised of organisms made of one cell with no nuclear membrane?

A. Monera.

B. Protista.

C. Fungi.

D. Algae.

A. Monera.

To answer this question, first note that algae are not a kingdom of their own. Some algae are in Monera, the kingdom that consists of unicellular prokaryotes with no true nucleus. Protista and fungi are both eukaryotic, with true nuclei, and are sometimes multicellular. Therefore, the **answer is (A)**.

63. Which of the following is found in the least abundance in organic molecules?

A. Phosphorus.

B. Potassium.

C. Carbon.

D. Oxygen.

B. Potassium.

Organic molecules consist mainly of carbon, hydrogen, and oxygen, with significant amounts of nitrogen, phosphorus, and often sulfur. Other elements, such as potassium, are present in much smaller quantities. Therefore, the **answer is (B)**. If you were not aware of this ranking, you might have been able to eliminate carbon and oxygen because of their prevalence, in any case.

64. Catalysts assist reactions by _______________

A. lowering effective activation energy.

B. maintaining precise pH levels.

C. keeping systems at equilibrium.

D. adjusting reaction speed.

A. Lowering effective activation energy.

Chemical reactions can be enhanced or accelerated by catalysts, which are present both with reactants and with products. They induce the formation of activated complexes, thereby lowering the effective activation energy—so that less energy is necessary for the reaction to begin. Although this often makes reactions faster, answer (D) is not as good a choice as the more generally applicable **answer (A)**, which is correct.

65. Accepted procedures for preparing solutions should be made with

A. alcohol.

B. hydrochloric acid.

C. distilled water.

D. tap water.

C. Distilled water.

Alcohol and hydrochloric acid should never be used to make solutions unless instructed to do so. All solutions should be made with distilled water as tap water contains dissolved particles which may affect the results of an experiment. The correct **answer is (C).**

66. Enzymes speed up reactions by ______________

A. utilizing ATP.

B. lowering pH, allowing reaction speed to increase.

C. increasing volume of substrate.

D. lowering energy of activation.

D. Lowering energy of activation.

Because enzymes are catalysts, they work the same way—they cause the formation of activated chemical complexes, which require a lower activation energy. Therefore, the **answer is (D).** ATP is an energy source for cells, and pH or volume changes may or may not affect reaction rate, so these answers can be eliminated.

67. When you step out of the shower, the floor feels colder on your feet than the bathmat. Which of the following is the correct explanation for this phenomenon?

A. The floor is colder than the bathmat.

B. Your feet have a chemical reaction with the floor, but not the bathmat.

C. Heat is conducted more easily into the floor.

D. Water is absorbed from your feet into the bathmat.

C. Heat is conducted more easily into the floor.

When you step out of the shower and onto a surface, the surface is most likely at room temperature, regardless of its composition (eliminating answer (A)). Your feet feel cold when heat is transferred from them to the surface, which happens more easily on a hard floor than a soft bathmat. This is because of differences in specific heat (the energy required to change temperature, which varies by material). Therefore, the **answer must be (C)**

68. Which of the following is NOT considered ethical behavior for a scientist?

A. Using unpublished data and citing the source.

B. Publishing data before other scientists have had a chance to replicate results.

C. Collaborating with other scientists from different laboratories.

D. Publishing work with an incomplete list of citations.

D. Publishing work with an incomplete list of citations.

One of the most important ethical principles for scientists is to cite all sources of data and analysis when publishing work. It is reasonable to use unpublished data (A), as long as the source is cited. Most science is published before other scientists replicate it (B), and frequently scientists collaborate with each other, in the same or different laboratories (C). These are all ethical choices. However, publishing work without the appropriate citations is unethical. Therefore, the **answer is (D).**

69. The chemical equation for water formation is: $2H_2 + O_2 \rightarrow 2H_2O$. Which of the following is an INCORRECT interpretation of this equation?

A. Two moles of hydrogen gas and one mole of oxygen gas combine to make two moles of water.

B. Two grams of hydrogen gas and one gram of oxygen gas combine to make two grams of water.

C. Two molecules of hydrogen gas and one molecule of oxygen gas combine to make two molecules of water.

D. Four atoms of hydrogen (combined as a diatomic gas) and two atoms of oxygen (combined as a diatomic gas) combine to make two molecules of water.

B. Two grams of hydrogen gas and one gram of oxygen gas combine to make two grams of water.

In any chemical equation, the coefficients indicate the relative proportions of molecules (or atoms), or of moles of molecules. They do not refer to mass, because chemicals combine in repeatable combinations of molar ratio (i.e. number of moles), but vary in mass per mole of material. Therefore, the answer must be the only choice that does not refer to numbers of particles, **answer (B)**, which refers to grams, a unit of mass.

70. Energy is measured with the same units as ______________

A. force.

B. momentum.

C. work.

D. power.

C. Work.

In SI units, energy is measured in Joules, i.e. (mass)(length squared)/(time squared). This is the same unit as is used for work. You can verify this by calculating that since work is force times distance, the units work out to be the same. Force is measured in Newtons in SI; momentum is measured in (mass)(length)/(time); power is measured in Watts (which equal Joules/second). Therefore, the **answer must be (C)**.

71. If the volume of a confined gas is increased, what happens to the pressure of the gas? You may assume that the gas behaves ideally, and that temperature and number of gas molecules remain constant.

A. The pressure increases.

B. The pressure decreases.

C. The pressure stays the same.

D. There is not enough information given to answer this question.

B. The pressure decreases.

Because we are told that the gas behaves ideally, you may assume that it follows the Ideal Gas Law, i.e. $PV = nRT$. This means that an increase in volume must be associated with a decrease in pressure (i.e. higher V means lower P), because we are also given that all the components of the right side of the equation remain constant. Therefore, the **answer must be (B)**.

72. A product of anaerobic respiration in animals is ______________

A. carbon dioxide.

B. lactic acid.

C. oxygen.

D. sodium chloride.

B. Lactic acid.

In animals, anaerobic respiration (i.e. respiration without the presence of oxygen) generates lactic acid as a byproduct. (Note that some anaerobic bacteria generate carbon dioxide from respiration of methane, and animals generate carbon dioxide in aerobic respiration.) Oxygen is not normally a by-product of respiration, though it is a product of photosynthesis, and sodium chloride is not strictly relevant in this question. Therefore, the **answer must be (B)**. By the way, lactic acid is believed to cause muscle soreness after anaerobic weight-lifting.

73. A Newton is fundamentally a measure of _______________

A. force.

B. momentum.

C. energy.

D. gravity.

A. Force.

In SI units, force is measured in Newtons. Momentum and energy each have different units, without equivalent dimensions. A Newton is one (kilogram)(meter)/(second squared), while momentum is measured in (kilgram)(meter)/(second) and energy, in Joules, is (kilogram)(meter squared)/(second squared). Although "gravity" can be interpreted as the force of gravity, i.e. measured in Newtons, fundamentally it is not required. Therefore, the **answer is (A)**.

74. Which change does NOT affect enzyme rate?

A. Increase the temperature.

B. Add more substrate.

C. Adjust the pH.

D. Use a larger cell.

D. Use a larger cell.

Temperature, chemical amounts, and pH can all affect enzyme rate. However, the chemical reactions take place on a small enough scale that the overall cell size is not relevant. Therefore, the **answer is (D)**.

75. Which of the following types of rock are made from magma?

A. Fossils.

B. Sedimentary.

C. Metamorphic.

D. Igneous.

D. Igneous.

Few fossils are found in metamorphic rock and virtually none found in igneous rocks. Igneous rocks are formed from magma and magma is so hot that any organisms trapped by it are destroyed. Metamorphic rocks are formed by high temperatures and great pressures. When fluid sediments are transformed into solid sedimentary rocks, the process is known as lithification. The **answer is (D).**

76. Which of the following is NOT an acceptable way for a student to acknowledge sources in a laboratory report?

A. The student tells his/her teacher what sources s/he used to write the report.

B. The student uses footnotes in the text, with sources cited, but not in correct MLA format.

C. The student uses endnotes in the text, with sources cited, in correct MLA format.

D. The student attaches a separate bibliography, noting each use of sources.

A. The student tells his/her teacher what sources s/he used to write the report.

It may seem obvious, but students are often unaware that scientists need to cite all sources used. For the young adolescent, it is not always necessary to use official MLA format (though this should be taught at some point). Students may properly cite references in many ways, but these references must be in writing, with the original assignment. Therefore, the **answer is (A).**

77. Animals with a notochord or a backbone are in the phylum

A. Arthropoda.

B. Chordata.

C. Mollusca.

D. Mammalia.

B. Chordata.

The phylum arthropoda contains spiders and insects and phylum mollusca contain snails and squid. Mammalia is a class in the phylum chordata. The **answer is (B).**

78. Which of the following is a correct explanation for scientific evolution'?

A. Giraffes need to reach higher for leaves to eat, so their necks stretch. The giraffe babies are then born with longer necks. Eventually, there are more long-necked giraffes in the population.

B. Giraffes with longer necks are able to reach more leaves, so they eat more and have more babies than other giraffes. Eventually, there are more long-necked giraffes in the population.

C. Giraffes want to reach higher for leaves to eat, so they release enzymes into their bloodstream, which in turn causes fetal development of longer-necked giraffes. Eventually, there are more long-necked giraffes in the population.

D. Giraffes with long necks are more attractive to other giraffes, so they get the best mating partners and have more babies. Eventually, there are more long-necked giraffes in the population.

B. Giraffes with longer necks are able to reach more leaves, so they eat more and have more babies than other giraffes. Eventually, there are more long-necked giraffes in the population.

Although evolution is often misunderstood, it occurs via natural selection. Organisms with a survival/reproductive advantage will produce more offspring. Over many generations, this changes the proportions of the population. In any case, it is impossible for a stretched neck (A) or a fervent desire (C) to result in a biologically mutated baby. Although there are traits that are naturally selected because of mate attractiveness and fitness (D), this is not the primary situation here, **so answer (B) is the best choice**.

79. Which of the following is a correct definition for 'chemical equilibrium'?

A. Chemical equilibrium is when the forward and backward reaction rates are equal. The reaction may continue to proceed forward and backward.

B. Chemical equilibrium is when the forward and backward reaction rates are equal, and equal to zero. The reaction does not continue.

C. Chemical equilibrium is when there are equal quantities of reactants and products.

D. Chemical equilibrium is when acids and bases neutralize each other fully.

A. Chemical equilibrium is when the forward and backward reaction rates are equal. The reaction may continue to proceed forward and backward.

Chemical equilibrium is defined as when the quantities of reactants and products are at a 'steady state' and are no longer shifting, but the reaction may still proceed forward and backward. The rate of forward reaction must equal the rate of backward reaction. Note that there may or may not be equal amounts of chemicals, and that this is not restricted to a completed reaction or to an acid-base reaction. Therefore, the **answer is (A)**.

80. Which of the following data sets is properly represented by a bar graph?

A. Number of people choosing to buy cars, vs. Color of car bought.

B. Number of people choosing to buy cars, vs. Age of car customer.

C. Number of people choosing to buy cars, vs. Distance from car lot to customer home.

D. Number of people choosing to buy cars, vs. Time since last car purchase.

A. Number of people choosing to buy cars, vs. Color of car bought.

A bar graph should be used only for data sets in which the independent variable is non-continuous (discrete), e.g. gender, color, etc. Any continuous independent variable (age, distance, time, etc.) should yield a scatter-plot when the dependent variable is plotted. Therefore, the **answer must be (A)**.

81. In a science experiment, a student needs to dispense very small measured amounts of liquid into a well-mixed solution. Which of the following is the best choice for his/her equipment to use?

A. Buret with Buret Stand, Stir-plate, Stirring Rod, Beaker.

B. Buret with Buret Stand, Stir-plate, Beaker.

C. Volumetric Flask, Dropper, Graduated Cylinder, Stirring Rod.

D. Beaker, Graduated Cylinder, Stir-plate.

B. Buret with Buret Stand, Stir-plate, Beaker.

The most accurate and convenient way to dispense small measured amounts of liquid in the laboratory is with a buret, on a buret stand. To keep a solution well-mixed, a magnetic stir-plate is the most sensible choice, and the solution will usually be mixed in a beaker. Although other combinations of materials could be used for this experiment, **choice (B)** is thus the simplest and best.

82. A laboratory balance is most appropriately used to measure the mass of which of the following?

A. Seven paper clips.

B. Three oranges.

C. Two hundred cells.

D. One student's elbow.

A. Seven paper clips.

Usually, laboratory/classroom balances can measure masses between approximately 0.01 gram and 100 grams. Therefore, answer (B) is too heavy and answer (C) is too light. Answer (D) is silly, but it is a reminder to instruct students not to lean on the balances or put their things near them. **Answer (A)**, which is likely to have a mass of a few grams, is correct in this case.

83. All of the following are measured in units of length, EXCEPT for:

A. Perimeter.

B. Distance.

C. Radius.

D. Area.

D. Area.

Perimeter is the distance around a shape; distance is equivalent to length; radius is the distance from the center (e.g. in a circle) to the edge. Area, however, is the squared-length-units measure of the size of a two-dimensional surface. Therefore, **the answer is (D)**.

84. What is specific gravity?

A. The mass of an object.

B. The ratio of the density of a substance to the density of water.

C. Density.

D. The pull of the earth's gravity on an object.

B. The ratio of the density of a substance to the density of water.

Mass is a measure of the amount of matter in an object. Density is the mass of a substance contained per unit of volume. Weight is the measure of the earth's pull of gravity on an object. The only option here is the ratio of the density of a substance to the density of water, **answer (B).**

85. What is the most accurate description of the Water Cycle?

A. Rain comes from clouds, filling the ocean. The water then evaporates and becomes clouds again.

B. Water circulates from rivers into groundwater and back, while water vapor circulates in the atmosphere.

C. Water is conserved except for chemical or nuclear reactions, and any drop of water could circulate through clouds, rain, ground-water, and surface-water.

D. Weather systems cause chemical reactions to break water into its atoms.

C. Water is conserved except for chemical or nuclear reactions, and any drop of water could circulate through clouds, rain, ground-water, and surface-water.

All natural chemical cycles, including the Water Cycle, depend on the principle of Conservation of Mass. (For water, unlike for elements such as nitrogen, chemical reactions may cause sources or sinks of water molecules.) Any drop of water may circulate through the hydrologic system, ending up in a cloud, as rain, or as surface or groundwater. Although answers (A) and (B) describe parts of the water cycle, the most comprehensive and correct **answer is (C)**.

86. The scientific name CANIS FAMILIARIS refers to the animal's ______________

A. kingdom and phylum.

B. genus and species.

C. class and species.

D. type and family.

B. Genus and species.

To answer this question, you must be aware that genus and species are the most specific way to identify an organism, and that usually the genus is capitalized and the species, immediately following, is not. Furthermore, it helps to recall that 'Canis' is the genus for dogs, or canines. Therefore, the **answer must be (B)**. If you did not remember these details, you might recall that there is no such kingdom as 'Canis,' and that there isn't a category 'type' in official taxonomy. This could eliminate answers (A) and (D).

87. Members of the same animal species ________________

A. look identical.

B. never adapt differently.

C. are able to reproduce with one another.

D. are found in the same location.

C. Are able to reproduce with one another.

Although members of the same animal species may look alike (A), adapt alike (B), or be found near one another (D), the only requirement is that they be able to reproduce with one another. This ability to reproduce within the group is considered the hallmark of a species. Therefore, the **answer is (C)**.

88. Which of the following is/are true about scientists?

I. Scientists usually work alone.
II. Scientists usually work with other scientists.
III. Scientists achieve more prestige from new discoveries than from replicating established results.
IV. Scientists keep records of their own work, but do not publish it for outside review.

A. I and IV only.

B. II only.

C. II and III only.

D. III and IV only.

C. II and III only.

In the scientific community, scientists nearly always work in teams, both within their institutions and across several institutions. This eliminates (I) and requires (II), leaving only **answer choices (B) and (C)**. Scientists do achieve greater prestige from new discoveries, so the answer must be (C). Note that scientists must publish their work in peer-reviewed journals, eliminating (IV) in any case.

89. What is necessary for ion diffusion to occur spontaneously?

A. Carrier proteins.

B. Energy from an outside source.

C. A concentration gradient.

D. Cell flagellae.

C. A concentration gradient.

Spontaneous diffusion occurs when random motion leads particles to increase entropy by equalizing concentrations. Particles tend to move into places of lower concentration. Therefore, a concentration gradient is required, and the **answer is (C)**. No proteins (A), outside energy (B), or flagellae (D) are required for this process.

90. All of the following are considered Newton's Laws EXCEPT for:

A. An object in motion will continue in motion unless acted upon by an outside force.

B. For every action force, there is an equal and opposite reaction force.

C. Nature abhors a vacuum.

D. Mass can be considered the ratio of force to acceleration.

C. Nature abhors a vacuum.

Newton's Laws include his law of inertia (an object in motion (or at rest) will stay in motion (or at rest) until acted upon by an outside force) (A), his law that (Force)=(Mass)(Acceleration) (D), and his equal and opposite reaction force law (B). Therefore, the **answer to this question is (C)**, because "Nature abhors a vacuum" is not one of these.

91. A cup of hot liquid and a cup of cold liquid are both sitting in a room at comfortable room temperature and humidity. Both cups are thin plastic. Which of the following is a true statement?

A. There will be fog on the outside of the hot liquid cup, and also fog on the outside of the cold liquid cup.

B. There will be fog on the outside of the hot liquid cup, but not on the cold liquid cup.

C. There will be fog on the outside of the cold liquid cup, but not on the hot liquid cup.

D. There will not be fog on the outside of either cup.

C. **There will be fog on the outside of the cold liquid cup, but not on the hot liquid cup.**

Fog forms on the outside of a cup when the contents of the cup are colder than the surrounding air, and the cup material is not a perfect insulator. This happens because the air surrounding the cup is cooled to a lower temperature than the ambient room, so it has a lower saturation point for water vapor. Although the humidity had been reasonable in the warmer air, when that air circulates near the colder region and cools, water condenses onto the cup's outside surface. This phenomenon is also visible when someone takes a hot shower, and the mirror gets foggy. The mirror surface is cooler than the ambient air, and provides a surface for water condensation. Furthermore, the same phenomenon is why defrosters on car windows send heat to the windows—the warmer window does not permit as much condensation. Therefore, the correct **answer is (C)**.

92. A ball rolls down a smooth hill. You may ignore air resistance. Which of the following is a true statement?

A. The ball has more energy at the start of its descent than just before it hits the bottom of the hill, because it is higher up at the beginning.

B. The ball has less energy at the start of its descent than just before it hits the bottom of the hill, because it is moving more quickly at the end.

C. The ball has the same energy throughout its descent, because positional energy is converted to energy of motion.

D. The ball has the same energy throughout its descent, because a single object (such as a ball) cannot gain or lose energy.

C. The ball has the same energy throughout its descent, because positional energy is converted to energy of motion.

The principle of Conservation of Energy states that (except in cases of nuclear reaction, when energy may be created or destroyed by conversion to mass), "Energy is neither created nor destroyed, but may be transformed." Answers (A) and (B) give you a hint in this question—it is true that the ball has more Potential Energy when it is higher, and that it has more Kinetic Energy when it is moving quickly at the bottom of its descent. However, the total sum of all kinds of energy in the ball remains constant, if we neglect 'losses' to heat/friction. Note that a single object can and does gain or lose energy when the energy is transferred to or from a different object. Conservation of Energy applies to systems, not to individual objects unless they are isolated. Therefore, the **answer must be (C)**.

93. A long silver bar has a temperature of 50 degrees Celsius at one end and 0 degrees Celsius at the other end. The bar will reach thermal equilibrium (barring outside influence) by the process of heat __________.

A. conduction.

B. convection.

C. radiation.

D. phase change.

A. conduction.

Heat conduction is the process of heat transfer via solid contact. The molecules in a warmer region vibrate more rapidly, jostling neighboring molecules and accelerating them. This is the dominant heat transfer process in a solid with no outside influences. Recall, also, that convection is heat transfer by way of fluid currents; radiation is heat transfer via electromagnetic waves; phase change can account for heat transfer in the form of shifts in matter phase. The answer to this question must **therefore be (A)**.

94. ____________ are cracks in the plates of the earth's crust, along which the plates move.

A. Faults

B. Ridges

C. Earthquakes

D. Volcanoes

A. Faults.

Faults are cracks in the earth's crust, and when the earth moves, an earthquake results. Faults may lead to mismatched edges of ground, forming ridges, and ground shape may also be determined by volcanoes. The answer to this question must **therefore be (A).**

95. Fossils are usually found in _____________ rock.

A. igneous.

B. sedimentary.

C. metamorphic.

D. cumulus.

B. Sedimentary

Fossils are formed by layers of dirt and sand settling around organisms, hardening, and taking an imprint of the organisms. When the organism decays, the hardened imprint is left behind. This is most likely to happen in rocks that form from layers of settling dirt and sand, i.e. sedimentary rock. Note that igneous rock is formed from molten rock from volcanoes (lava), while metamorphic rock can be formed from any rock under very high temperature and pressure changes. 'Cumulus' is a descriptor for clouds, not rocks. The best answer is **therefore (B)**.

96. Which of the following is NOT a common type of acid in 'acid rain' or acidified surface water?

A. Nitric acid.

B. Sulfuric acid.

C. Carbonic acid.

D. Hydrofluoric acid.

D. Hydrofluoric acid.

Acid rain forms predominantly from pollutant oxides in the air (usually nitrogen-based NO_x or sulfur-based SO_x), which become hydrated into their acids (nitric or sulfuric acid). Because of increased levels of carbon dioxide pollution, carbonic acid is also common in acidified surface water environments. Hydrofluoric acid can be found, but it is much less common. In general, carbon, nitrogen, and sulfur are much more prevalent in the environment than fluorine. Therefore, the **answer is (D)**.

97. Which of the following is NOT true about phase change in matter?

A. Solid water and liquid ice can coexist at water's freezing point.

B. At 7 degrees Celsius, water is always in liquid phase.

C. Matter changes phase when enough energy is gained or lost.

D. Different phases of matter are characterized by differences in molecular motion.

B. At 7 degrees Celsius, water is always in liquid phase.

According to the molecular theory of matter, molecular motion determines the 'phase' of the matter, and the energy in the matter determines the speed of molecular motion. Solids have vibrating molecules that are in fixed relative positions; liquids have faster molecular motion than their solid forms, and the molecules may move more freely but must still be in contact with one another; gases have even more energy and more molecular motion. (Other phases, such as plasma, are yet more energetic.) At the 'freezing point' or 'boiling point' of a substance, both relevant phases may be present. For instance, water at zero degrees Celsius may be composed of some liquid and some solid, or all liquid, or all solid. Pressure changes, in addition to temperature changes, can cause phase changes. For example, nitrogen can be liquefied under high pressure, even though its boiling temperature is very low. Therefore, the **correct answer must be (B)**. Water may be a liquid at that temperature, but it may also be a solid, depending on ambient pressure.

98. Which of the following is the longest (largest) unit of geological time?

A. Solar Year.

B. Epoch.

C. Period.

D. Era.

D. Era.

Geological time is measured by many units, but the longest unit listed here (and indeed the longest used to describe the biological development of the planet) is the Era. Eras are subdivided into Periods, which are further divided into Epochs. Therefore, the **answer is (D).**

99. Extensive use of antibacterial soap has been found to increase the virulence of certain infections in hospitals. Which of the following might be an explanation for this phenomenon?

A. Antibacterial soaps do not kill viruses.

B. Antibacterial soaps do not incorporate the same antibiotics used as medicine.

C. Antibacterial soaps kill a lot of bacteria, and only the hardiest ones survive to reproduce.

D. Antibacterial soaps can be very drying to the skin.

C. **Antibacterial soaps kill a lot of bacteria, and only the hardiest ones survive to reproduce**.

All of the answer choices in this question are true statements, but the question specifically asks for a cause of increased disease virulence in hospitals. This phenomenon is due to natural selection. The bacteria that can survive contact with antibacterial soap are the strongest ones, and without other bacteria competing for resources, they have more opportunity to flourish. This problem has led to several antibiotic-resistant bacterial diseases in hospitals nation-wide. Therefore, the **answer is (C)**. However, note that answers (A) and (D) may be additional problems with over-reliance on antibacterial products.

100. Which of the following is a correct explanation for astronaut 'weightlessness'?

A. Astronauts continue to feel the pull of gravity in space, but they are so far from planets that the force is small.

B. Astronauts continue to feel the pull of gravity in space, but spacecraft have such powerful engines that those forces dominate, reducing effective weight.

C. Astronauts do not feel the pull of gravity in space, because space is a vacuum.

D. Astronauts do not feel the pull of gravity in space, because black hole forces dominate the force field, reducing their masses.

A. Astronauts continue to feel the pull of gravity in space, but they are so far from planets that the force is small.

Gravity acts over tremendous distances in space (theoretically, infinite distance, though certainly at least as far as any astronaut has traveled). However, gravitational force is inversely proportional to distance squared from a massive body. This means that when an astronaut is in space, s/he is far enough from the center of mass of any planet that the gravitational force is very small, and s/he feels 'weightless'. Space is mostly empty (i.e. vacuum), and there are some black holes, and spacecraft do have powerful engines. However, none of these has the effect attributed to it in the incorrect answer choices (B), (C), or (D). The answer to this question must **therefore be (A).**

101. The theory of 'sea floor spreading' explains ______________

A. the shapes of the continents.

B. how continents collide.

C. how continents move apart.

D. how continents sink to become part of the ocean floor.

C. How continents move apart.

In the theory of 'sea floor spreading', the movement of the ocean floor causes continents to spread apart from one another. This occurs because crust plates split apart, and new material is added to the plate edges. This process pulls the continents apart, or may create new separations, and is believed to have caused the formation of the Atlantic Ocean. The **answer is (C)**.

102. Which of the following animals are most likely to live in a tropical rainforest?

A. Reindeer.

B. Monkeys.

C. Puffins.

D. Bears.

B. Monkeys.

The tropical rain forest biome is hot and humid, and is very fertile—it is thought to contain almost half of the world's species. Reindeer (A), puffins (C), and bears (D), however, are usually found in much colder climates. There are several species of monkeys that thrive in hot, humid climates, so **answer (B) is correct.**

103. Which of the following is NOT a type of volcano?

A. Shield Volcanoes.

B. Composite Volcanoes.

C. Stratus Volcanoes.

D. Cinder Cone Volcanoes.

C. Stratus Volcanoes.

There are three types of volcanoes. Shield volcanoes (A) are associated with non-violent eruptions and repeated lava flow over time. Composite volcanoes (B) are built from both lava flow and layers of ash and cinders. Cinder cone volcanoes (D) are associated with violent eruptions, such that lava is thrown into the air and becomes ash or cinder before falling and accumulating. **'Stratus' (C)** is a type of cloud, not volcano, so it is the correct answer to this question.

104. Which of the following is NOT a property of metalloids?

A. Metalloids are solids at standard temperature and pressure.

B. Metalloids can conduct electricity to a limited extent.

C. Metalloids are found in groups 13 through 17.

D. Metalloids all favor ionic bonding.

D. **Metalloids all favor ionic bonding**.

Metalloids are substances that have characteristics of both metals and nonmetals, including limited conduction of electricity and solid phase at standard temperature and pressure. Metalloids are found in a 'stair-step' pattern from Boron in group 13 through Astatine in group 17. Some metalloids, e.g. Silicon, favor covalent bonding. Others, e.g. Astatine, can bond ionically. Therefore, **the answer is (D).** Recall that metals/nonmetals/metalloids are not strictly defined by Periodic Table group, so their bonding is unlikely to be consistent with one another.

105. Which of these is a true statement about loamy soil?

A. Loamy soil is gritty and porous.

B. Loamy soil is smooth and a good barrier to water.

C. Loamy soil is hostile to microorganisms.

D. Loamy soil is velvety and clumpy.

D. **Loamy soil is velvety and clumpy**.

The three classes of soil by texture are: Sandy (gritty and porous), Clay (smooth, greasy, and most impervious to water), and Loamy (velvety, clumpy, and able to hold water and let water flow through). In addition, loamy soils are often the most fertile soils. Therefore, the **answer must be (D)**.

106. Lithification refers to the process by which unconsolidated sediments are transformed into _______________

A. metamorphic rocks.

B. sedimentary rocks.

C. igneous rocks.

D. lithium oxide.

B. Sedimentary rocks.

Lithification is the process of sediments coming together to form rocks, i.e. sedimentary rock formation. Metamorphic and igneous rocks are formed via other processes (heat and pressure or volcano, respectively). Lithium oxide shares a word root with 'lithification' but is otherwise unrelated to this question. Therefore, the **answer must be (B)**.

107. Igneous rocks can be classified according to which of the following?

A. Texture.

B. Composition.

C. Formation process.

D. All of the above.

D. All of the above.

Igneous rocks, which form from the crystallization of molten lava, are classified according to many of their characteristics, including texture, composition, and how they were formed. Therefore, **the answer is (D).**

108. Which of the following is the most accurate definition of a nonrenewable resource?

A. A nonrenewable resource is never replaced once used.

B. A nonrenewable resource is replaced on a timescale that is very long relative to human life-spans.

C. A nonrenewable resource is a resource that can only be manufactured by humans.

D. A nonrenewable resource is a species that has already become extinct.

B. A nonrenewable resource is replaced on a timescale that is very long relative to human life-spans.

Renewable resources are those that are renewed, or replaced, in time for humans to use more of them. Examples include fast-growing plants, animals, or oxygen gas. (Note that while sunlight is often considered a renewable resource, it is actually a nonrenewable but extremely abundant resource.) Nonrenewable resources are those that renew themselves only on very long timescales, usually geologic timescales. Examples include minerals, metals, or fossil fuels. Therefore, the **correct answer is (B)**.

109. The theory of 'continental drift' is supported by which of the following?

A. The way the shapes of South America and Europe fit together.

B. The way the shapes of Europe and Asia fit together.

C. The way the shapes of South America and Africa fit together.

D. The way the shapes of North America and Antarctica fit together.

C. The way the shapes of South America and Africa fit together.

The theory of 'continental drift' states that many years ago, there was one land mass on the earth ('pangea'). This land mass broke apart via earth crust motion, and the continents drifted apart as separate pieces. This is supported by the shapes of South America and Africa, which seem to fit together like puzzle pieces if you look at a globe. Note that answer choices (A), (B), and (D) give either land masses that do not fit together, or those that are still attached to each other. Therefore, the **answer must be (C)**.

110. When water falls to a cave floor and evaporates, it may deposit calcium carbonate. This process leads to the formation of which of the following?

A. Stalactites.

B. Stalagmites.

C. Fault lines.

D. Sedimentary rocks.

B. Stalagmites.

To answer this question, recall the trick to remember the kinds of crystals formed in caves. Stalactites have a 'T' in them, because they form hanging from the ceiling (resembling a 'T'). Stalagmites have an 'M' in them, because they make bumps on the floor (resembling an 'M'). Note that fault lines and sedimentary rocks are irrelevant to this question. Therefore, **the answer must be (B)**.

111. A child has type O blood. Her father has type A blood, and her mother has type B blood. What are the genotypes of the father and mother, respectively?

A. AO and BO.

B. AA and AB.

C. OO and BO.

D. AO and BB.

A. AO and BO.

Because O blood is recessive, the child must have inherited two O's—one from each of her parents. Since her father has type A blood, his genotype must be AO; likewise her mother's blood must be BO. Therefore, only **answer (A)** can be correct.

112. Which of the following is the best definition for 'meteorite'?

A. A meteorite is a mineral composed of mica and feldspar.

B. A meteorite is material from outer space, that has struck the earth's surface.

C. A meteorite is an element that has properties of both metals and nonmetals.

D. A meteorite is a very small unit of length measurement.

B. A meteorite is material from outer space, that has struck the earth's surface.

Meteoroids are pieces of matter in space, composed of particles of rock and metal. If a meteoroid travels through the earth's atmosphere, friction causes burning and a "shooting star"—i.e. a meteor. If the meteor strikes the earth's surface, it is known as a meteorite. Note that although the suffix –ite often means a mineral, answer (A) is incorrect. Answer (C) refers to a 'metalloid' rather than a 'meteorite', and answer (D) is simply a misleading pun on 'meter'. Therefore, the **answer is (B)**.

113.A white flower is crossed with a red flower. Which of the following is a sign of incomplete dominance?

A. Pink flowers.

B. Red flowers.

C. White flowers.

D. No flowers.

A. Pink flowers.

Incomplete dominance means that neither the red nor the white gene is strong enough to suppress the other. Therefore both are expressed, leading in this case to the formation of pink flowers. Therefore, the **answer is (A)**.

114. What is the source for most of the United States' drinking water?

A. Desalinated ocean water.

B. Surface water (lakes, streams, mountain runoff).

C. Rainfall into municipal reservoirs.

D. Groundwater.

D. Groundwater.

Groundwater currently provides drinking water for 53% of the population of the United States. (Although groundwater is often less polluted than surface water, it can be contaminated and it is very hard to clean once it is polluted. If too much groundwater is used from one area, then the ground may sink or shift, or local salt water may intrude from ocean boundaries.) The other answer choices can be used for drinking water, but they are not the most widely used. Therefore, **the answer is (D)**.

115. Which is the correct sequence of insect development?

A. Egg, pupa, larva, adult.

B. Egg, larva, pupa, adult.

C. Egg, adult, larva, pupa.

D. Pupa, egg, larva, adult.

B. Egg, larva, pupa, adult.

An insect begins as an egg, hatches into a larva (caterpillar), forms a pupa (cocoon), and emerges as an adult (moth). Therefore, the **answer is (B)**.

116. A wrasse (fish) cleans the teeth of other fish by eating away plaque. This is an example of ____________________ between the fish.

A. parasitism.

B. symbiosis (mutualism).

C. competition.

D. predation.

B. Symbiosis (mutualism).

When both species benefit from their interaction in their habitat, this is called 'symbiosis', or 'mutualism'. In this example, the wrasse benefits from having a source of food, and the other fish benefit by having healthier teeth. Note that 'parasitism' is when one species benefits at the expense of the other, 'competition' is when two species compete with one another for the same habitat or food, and 'predation' is when one species feeds on another. Therefore, the **answer is (B)**.

117. What is the main obstacle to using nuclear fusion for obtaining electricity?

A. Nuclear fusion produces much more pollution than nuclear fission.

B. There is no obstacle; most power plants us nuclear fusion today.

C. Nuclear fusion requires very high temperature and activation energy.

D. The fuel for nuclear fusion is extremely expensive.

C. Nuclear fusion requires very high temperature and activation energy.

Nuclear fission is the usual process for power generation in nuclear power plants. This is carried out by splitting nuclei to release energy. The sun's energy is generated by nuclear fusion, i.e. combination of smaller nuclei into a larger nucleus. Fusion creates much less radioactive waste, but it requires extremely high temperature and activation energy, so it is not yet feasible for electricity generation. Therefore, the **answer is (C)**.

118. Which of the following is a true statement about radiation exposure and air travel?

A. Air travel exposes humans to radiation, but the level is not significant for most people.

B. Air travel exposes humans to so much radiation that it is recommended as a method of cancer treatment.

C. Air travel does not expose humans to radiation.

D. Air travel may or may not expose humans to radiation, but it has not yet been determined.

A. Air travel exposes humans to radiation, but the level is not significant for most people.

Humans are exposed to background radiation from the ground and in the atmosphere, but these levels are not considered hazardous under most circumstances, and these levels have been studied extensively. Air travel does create more exposure to atmospheric radiation, though this is much less than people usually experience through dental X-rays or other medical treatment. People whose jobs or lifestyles include a great deal of air flight may be at increased risk for certain cancers from excessive radiation exposure. Therefore, the **answer is (A)**.

119. Which process(es) result(s) in a haploid chromosome number?

A. Mitosis.

B. Meiosis.

C. Both mitosis and meiosis.

D. Neither mitosis nor meiosis.

B. Meiosis.

Meiosis is the division of sex cells. The resulting chromosome number is half the number of parent cells, i.e. a 'haploid chromosome number'. Mitosis, however, is the division of other cells, in which the chromosome number is the same as the parent cell chromosome number. Therefore, the **answer is (B)**.

120. Which of the following is NOT a member of Kingdom Fungi?

A. Mold.

B. Blue-green algae.

C. Mildew.

D. Mushrooms.

B. Blue-green Algae.

Mold (A), mildew (C), and mushrooms (D) are all types of fungus. Blue-green algae, however, is in Kingdom Monera. Therefore, the **answer is (B)**.

121. Which of the following organisms use spores to reproduce?

A. Fish.

B. Flowering plants.

C. Conifers.

D. Ferns.

D. **Ferns.**

Ferns, in Division Pterophyta, reproduce with spores and flagellated sperm. Flowering plants reproduce via seeds, and conifers reproduce via seeds protected in cones (e.g. pinecone). Fish, of course, reproduce sexually. Therefore, the **answer is (D)**.

122. What is the main difference between the 'condensation hypothesis' and the 'tidal hypothesis' for the origin of the solar system?

A. The tidal hypothesis can be tested, but the condensation hypothesis cannot.

B. The tidal hypothesis proposes a near collision of two stars pulling on each other, but the condensation hypothesis proposes condensation of rotating clouds of dust and gas.

C. The tidal hypothesis explains how tides began on planets such as Earth, but the condensation hypothesis explains how water vapor became liquid on Earth.

D. The tidal hypothesis is based on Aristotelian physics, but the condensation hypothesis is based on Newtonian mechanics.

B. **The tidal hypothesis proposes a near collision of two stars pulling on each other, but the condensation hypothesis proposes condensation of rotating clouds of dust and gas.**

Most scientists believe the 'condensation hypothesis,' i.e. that the solar system began when rotating clouds of dust and gas condensed into the sun and planets. A minority opinion is the 'tidal hypothesis,' i.e. that the sun almost collided with a large star. The large star's gravitational field would have then pulled gases out of the sun; these gases are thought to have begun to orbit the sun and condense into planets. Because both of these hypotheses deal with ancient, unrepeatable events, neither can be tested, eliminating answer (A). Note that both 'tidal' and 'condensation' have additional meanings in physics, but those are not relevant here, eliminating answer (C). Both hypotheses are based on best guesses using modern physics, eliminating answer (D). Therefore, the **answer is (B)**.

123. Which of the following units is NOT a measure of distance?

A. AU (astronomical unit).

B. Light year.

C. Parsec.

D. Lunar year.

D. Lunar year.

Although the terminology is sometimes confusing, it is important to remember that a 'light year' (B) refers to the distance that light can travel in a year. Astronomical Units (AU) (A) also measure distance, and one AU is the distance between the sun and the earth. Parsecs (C) also measure distance, and are used in astronomical measurement- they are very large, and are usually used to measure interstellar distances. A lunar year, or any other kind of year for a planet or moon, is the *time* measure of that body's orbit. Therefore, the answer to this **question is (D)**.

124. The salinity of ocean water is closest to ____________ .

A. 0.035 %

B. 0.35 %

C. 3.5 %

D. 35 %

C. 3.5 %

Salinity, or concentration of dissolved salt, can be measured in mass ratio (i.e. mass of salt divided by mass of sea water). For Earth's oceans, the salinity is approximately 3.5 %, or 35 parts per thousand. Note that answers (A) and (D) can be eliminated, because (A) is so dilute as to be hardly saline, while (D) is so concentrated that it would not support ocean life. Therefore, the **answer is (C)**.

125. Which of the following will not change in a chemical reaction?

A. Number of moles of products.

B. Atomic number of one of the reactants.

C. Mass (in grams) of one of the reactants.

D. Rate of reaction.

B. Atomic number of one of the reactants.

Atomic number, i.e. the number of protons in a given element, is constant unless involved in a nuclear reaction. Meanwhile, the amounts (measured in moles (A) or in grams(C)) of reactants and products change over the course of a chemical reaction, and the rate of a chemical reaction (D) may change due to internal or external processes. Therefore, the **answer is (B)**.

XAMonline, INC. 21 Orient Ave. Melrose, MA 02176
Toll Free number 800-509-4128
TO ORDER Fax 781-662-9268 OR www.XAMonline.com

NEW MEXICO TEACHER ASSESSMENT - NMTA - 2007

P0# Store/School:

Address 1:

Address 2 (Ship to other):

City, State Zip

Credit card number_______-________-_________-________ expiration_______

EMAIL ______________________________

PHONE FAX

13# ISBN 2007	TITLE	Qty	Retail	Total
978-1-58197-750-9	NMTA New Mexico Assessment of Teacher Basic Skills 01			
978-1-58197-751-6	NMTA New Mexico Assessment of Teacher Competency 03, 04 , 05			
978-1-58197-752-3	NMTA Elementary Education 11			
978-1-58197-754-7	NMTA Language Arts 12			
978-1-58197-760-8	NMTA Reading 13			
978-1-58197-755-4	NMTA Mathematics 14			
978-1-58197-761-5	NMTA Science 15			
978-1-58197-762-2	NMTA History, Geography, Civics, and Government 16			
978-1-58197-753-0	NMTA French Sample Test 18			
978-1-58197-763-9	NMTA Spanish 20			
978-1-58197-764-6	NMTA Visual Arts Sample Test 22			
978-1-58197-756-1	NMTA Middle Level Language Arts 23			
978-1-58197-757-8	NMTA Middle Level Mathematics 24			
978-1-58197-758-5	NMTA Middle Level Science 25			
978-1-58197-759-2	NMTA Middle Level Social Studies 26			
			SUBTOTAL	
			Ship	$8.25
			TOTAL	

CPSIA information can be obtained at www.ICGtesting.com
Printed in the USA
LVOW09s0339250214

374982LV00002B/52/A